MW01598583

for Train-the-Trainer Programs: Truck-Driving Training

SANDY SMITH

THOMSON

DELMAR LEARNING™

Australia Canada Mexico Singapore Spain United Kingdom United States

THOMSON
DELMAR LEARNING

Train the Trainer Student Course Book
PTDI/TCA

Vice President, Technology and Trades SBU:
Alar Elken

Editorial Director:
Sandy Clark

Product Development Manager:
Tim Waters

Editorial Assistant:
Kristen Shenfield

Marketing Director:
Cyndi Eichelman

Channel Manager:
Beth A. Lutz

Marketing Specialist:
Brian McGrath

Production Director:
Mary Ellen Black

Production Manager:
Andrew Crouth

Production Editor:
Sharon Popson

Library of Congress Cataloging-in-Publication Data:
ISBN: 1401805124

NOTICE TO THE READER

Contents

Preface

The purpose of the Delmar *Train-the-Trainer Program Driver Trainer Student Handbook* is to offer teaching methodology, learning philosophy, and professional development for aspiring trainers of truck-driving students. It is intended to be used by schools and training organizations to educate driver trainers in all aspects of training student drivers. It is designed to work effectively in an instructor-led classroom environment, but it can also be utilized in a self-study or customized learning environment.

WHY THIS TEXT WAS DEVELOPED

The field of truck-driving training has undergone drastic changes in the past decade and it continues to change. This text is designed to respond to the industry-wide need for quality training materials to effectively train driver trainers.

ORGANIZATION OF MATERIAL

Written in a clear, easy-to-read style, this interactive handbook is user friendly and can serve as a resource that learners and educators can draw on to develop their own plans of action for learning in the classroom. Six sections of the text focus on teaching methods, learning techniques, and professional career development. Chapter 1 includes information on the reasons people choose to become driver trainers and the career opportunities available to a person entering the the truck-driving training industry. This section also contains information about the critical elements in an effective driver trainer program, as well as the responsibilities of an effective driver trainer. Chapter 8 contains information on professional development, including issues related to finding and maintaining satisfactory employment in this field.

HOW TO USE THIS MATERIAL

This text provides background information and integration of theory and applied methods for driver trainers. It is designed as a resource that learners and educators can use to develop their own plans of action for training students in the classroom, on the range, and on the road. It is a resource for trainers to use when preparing their own lessons to teach experienced drivers to become effective instructors.

ACKNOWLEDGMENTS

Partnership with PTDI Standards for the Truck-Driving Training Industry

Eric Rice for his original work on this project

Introduction and Purpose

The purpose of this student handbook is to help prepare trainers to teach new truck drivers. It is part of a series of train-the-trainer materials designed to improve the effectiveness of training in the trucking industry. The series includes an *Instructor's Guide* for driver trainers to use and videos that provide hands-on, interactive opportunities to develop training skills.

Driver trainers need to be more than just good drivers; they need to understand the technical subjects and master the technical skills they want their students to acquire and demonstrate. In addition, they must also possess and use specific pedagogical and communication skills to successfully facilitate student learning. This handbook is designed to help those who train truck-driving trainers to become effective instructors.

Students entering truck-driving training schools expect and deserve the best possible education you can provide. These adult learners have chosen your school for their education and you as the teachers to educate them. A truck-driving training instructor's primary focus is to provide education and training that will equip students with the necessary skills and abilities to be professional, entry-level truck drivers. Educators constantly strive to advance and develop the standards of education and instruction offered in their schools. Obtaining that goal can be challenging, but with the right tools, it is a challenge that can be successfully met.

The purpose of the handbook is to aid educators in meeting the objective of advancing and improving the standards of education in your school.

HOW THE HANDBOOK IS ORGANIZED

Learners and educators can use this handbook to develop their own plans of action for training students in the classroom, on the range, and on the road. It is a resource for trainers to use when preparing their own lessons to teach experienced drivers to become effective instructors. Several chapters directly relate to the skills that experienced trainers have said are the most important skills driver trainers need to be effective teachers. In addition to information on the characteristics of the adult learner, learner diversity, curriculum development, classroom management, presentation skills, and student assessment, the handbook contains information on professional development in the truck-driving field.

The handbook contains information that focuses on the most important topics to keep in mind when creating effective driver trainer programs.

CHAPTER 1

Why Become a Driver Trainer?

OVERVIEW

Welcome to the rewarding world of teaching. Making the decision to pursue your career as a professional driver trainer is an important one. As an educator, you will be responsible for training others by formal instruction and supervised practice, as well as helping to develop the characteristics necessary to equip your students to become professional truck drivers. As an educator in this field, your primary focus is to provide education and training in truck driving and related areas. The training you provide will equip your students with the skills, abilities, and knowledge they will need to be competitive in today's dynamic job market. The materials in the following chapter provide background information important to consider concerning your decision to become a driver trainer.

OBJECTIVES

As a result of reading and studying this chapter, you should be able to:

- Identify the reasons you have chosen to become a driver trainer.

- Understand the career opportunities available to a person entering the education field in the truck-driving training industry.

- Understand the critical elements in an effective driver-trainer program.

- Understand the responsibilities and characteristics of an effective driver trainer.

INTRODUCTION

Your decision to become a driver trainer gives you an important opportunity to affect the lives of countless students each year as they pursue their career goals. Professional truck driving is a demanding and responsible career. Competent, safe, skilled, responsible drivers are necessary to ensure safe highways. Professional drivers who can make independent judgments and are accountable for their actions are an asset to their companies and a credit to their profession. Thousands of professional truck drivers are seen doing their jobs each day by the motoring public as they move goods across the nation. These drivers represent the image the public has of this profession. They know they must perform better than the average driver in order to maintain a positive image for their industry. Your role in providing good training and education to these drivers is important in maintaining the quality standards that are crucial to this industry.

In addition to maintaining a professional image, motor carriers want drivers who perform their jobs in a safe manner because they entrust valuable vehicles and cargo to their care. These drivers also represent the company's corporate image to the public. As a driver trainer, you will be entrusted with helping to maintain this image as a result of the educational quality standards you maintain and the competent students you train and graduate.

Many more new truck drivers are needed each year than graduate from the nation's public and private truck-driving schools. Besides helping to ensure that enough high-quality truck drivers are produced to meet the increasing demand, as a driver trainer you will also be in a unique position to advance the standards of education and instruction you provide in your classroom.

CAREER OPPORTUNITIES

Formal training is the most reliable way to learn the special skills required for truck driving. Formal, supervised training is available for students interested in entering this field from private truck-driver training schools, public truck-driver training institutions, and in-house motor carrier training programs. Each of these organizations requires effective driver trainers in order to successfully graduate student trainees, which means there are many opportunities for interested people to pursue the career of a driver trainer.

Driver trainers must possess a combination of education and experience that qualifies them to educate truck drivers. In addition to the mastery of technical skills related to truck driving, driver trainers must also possess teaching skills that enable student learning and achievement of job-related competencies.

Many schools provide training according to industry-developed skill standards that were created through the collaboration of many trucking industry stakeholders. Skill standards represent the benchmarks necessary for proficiencies in various fields. These skill standards have helped the truck-driving training industry identify curriculum standards based on the skills, knowledge, tasks, and duties required of entry-level truck drivers. They help to drive consistency of training across the industry and ensure that truck-driving schools produce graduates who can meet the performance demands of this field.

Regardless of the school or training organization in which you choose to work, the primary objective will be the same: *to train competent, qualified professional truck drivers for successful employment in their chosen field.* As a trainer in the truck-driving training field you will need to create the correct mix of self-motivation, sincerity, excellent technical skills, effective communication skills, strong work ethic, enthusiasm, safety consciousness, professional image, and a strong personal desire to learn, grow, and succeed in order to produce qualified trainees. In addition to producing professional truck drivers, effective driver trainers must instill a desire in their students to continually improve on their skills and to make a commitment to lifelong learning in order to meet the changing demands of this profession.

VISION OF TEACHING

Educators must develop a vision for the future of their profession and incorporate that vision into their teaching. To develop a vision, teachers must make a sincere commitment to the personal and professional growth of their students. Effective driver trainers must also be aware of industry trends and relate those trends to the skill-development needs of their students. In addition to incorporating a personal vision, driver trainers should consider all the elements necessary to create a learning environment that will produce graduates who can meet industry performance requirements. Stakeholders in the truck-driving training field have identified factors considered critical for success in training truck-driving students. These elements include:

❑ The quality and safety of training equipment

❑ The curriculum content of the training program

- ❏ The amount of time students will have access to training equipment

- ❏ The ratio of instructor to students

- ❏ Specific lesson plans for each session and skill area

- ❏ The ability of students to practice skills under realistic conditions

- ❏ Methods to evaluate program outcomes, including assessment of student knowledge, skills, and proficiencies

Each of these elements must be addressed to develop comprehensive and high-quality truck-driving training programs.

INDUSTRY SKILL NEEDS

The transportation industry has been increasingly affected by a shortage of skilled workers. As an example, the skill shortages within the long-haul trucking industry reflect the statistics across the transportation industry: Almost 3 million drivers move more than $38 billion worth of products annually across America. The vitality of this and other transportation industries depends on attracting and retaining new sources of skilled workers. A Gallup study estimated an annual need for 80,000 additional skilled truck drivers in the workforce due to both industry growth and attrition. The U.S. Bureau of Labor Statistics estimates that the number of drivers needed to meet demand will increase more than 10 percent throughout the current decade. At the same time, more than 500,000 drivers are expected to retire or change occupations. These facts underline the continuing need for skilled drivers, and for trained individuals to teach these new drivers. These statistics indicate that the demand for effective driver trainers is anticipated to remain at the same high levels as the demand for drivers, thereby producing continuing career opportunities for new driver trainers.

CHARACTERISTICS OF AN EFFECTIVE TRAINER

The basic function of a driver trainer is to educate truck-driving students. Teaching is an intellectual exercise that requires the ability to invent, adapt, and create new techniques and procedures to meet the changing demands of learners, as well as the trucking industry. To be effective in the role of teacher requires certain qualities, characteristics, traits, skills, and techniques. The qualities of an effective teacher are a composite of various roles and characteristics necessary to achieve the goals of success with learners.

The profile of a successful educator includes the ability to function in many roles, as student needs dictate. Successful driver trainers will find themselves at times being a mentor, coach, motivator, friend, disciplinarian, negotiator, nurturer, and entertainer. The most successful vocational training takes place in classrooms and laboratory settings that replicate the expectations and demands of a business environment. It is important to model to your students what they will encounter on the job, and setting the right tone in the training classroom can help to support the development of acceptable workplace performance and behaviors.

The classroom is a place where adult learners interact continually through verbal and nonverbal communication. The educator who is knowledgeable on a number of topics will be a more interesting role model to students. If you as a driver trainer can converse with your students on a variety of topics, this will help you be an effective educator.

An effective driver trainer must possess a high level of expertise in the technical skills he or she teaches. Educators are judged by their ability to perform and demonstrate technical skills. Students will admire and respect the technical and teaching abilities of trainers who demonstrate mastery of their subject matter. In doing so, trainers will earn the admiration of their learners while imparting critical information that will help their students improve their own skills, knowledge, attitudes, understanding, and habits.

Education is a continuing, lifelong process and changes in the truck-driving field take place daily. It is in the best interests of the driver trainer to keep up with the field and demonstrate continuing competency in understanding and addressing changing trends in the curriculum and in practice. As a driver trainer, your commitment to personal career and professional development is a quality desired by employers that can be modeled for your students.

Your career development as a driver trainer does not end with the technical and theoretical knowledge of the skills you teach. You must also continue to develop your abilities and expertise as an educator. For example, managing your time strategically is an important personal characteristic of successful teachers. The key concepts of effective time management include *setting goals, establishing objectives, identifying priorities, analyzing your use of time, planning your time,* and *achieving a balance between work and your personal life, or down time.*

To influence the thoughts, opinions, and behaviors of your learners and to achieve recognition as an authority among your students, you must first have good self-esteem and self-confidence. If, as an instructor, you engage in systematic, purposeful action that is consistent with your values and goals, you will feel positive about your abilities and will exhibit confidence. Displaying this level of confidence will help to establish your authority with students. When your authority has been established, you will be able to maintain order in class and encourage a strong desire among your students to achieve success in their career, as you have in yours.

High-quality truck-driving training schools look for additional characteristics in their instructors. Employers seek trainers and instructors whose ethics and principles are solid, who are dependable and flexible, and who can function as cooperative team members. As a professional truck-driving trainer, you will be expected to show the initiative to start and follow through on tasks and to work independently, with little close supervision. In addition, educators must also exercise self-control and patience when dealing with challenging students. Adult learners require respectful and courteous treatment in order to make the most of their learning experience, and you will be responsible for setting a positive tone in your classroom to facilitate this process.

An effective driver trainer develops and projects self-confidence and models these attributes for students. Following are suggestions to help build self-confidence.

Self-Confidence Builders

❑ **Learn to like and accept yourself unconditionally.** Trust your instincts and consider yourself a valuable and worthwhile human being. If you develop and follow your principles and exhibit purposeful action that follows these beliefs, you will exhibit self-confidence and learn to trust your ability to say and do the right thing at the right time.

❑ **Be true to yourself.** The more you understand yourself and your motivations, the more likely you are to be self-assured and to remain confident in your abilities.

❑ **Understand your values.** The more you understand what you believe in and allow your actions to follow your values and beliefs, the more accepting you will be of yourself and others.

❑ **Be courageous.** Learn to accept yourself as who and what you are, not what others expect you to be. Learn to take risks and to find the courage to follow through on your actions and principles.

❑ **Recognize that you are unique and have special gifts.** Learn to make the most of the unique skills, talents, abilities, and insights you possess. Incorporate these unique features into your personality and your attitude toward career development.

❏ **Work toward goal achievement.** Set achievable goals for yourself in obtainable increments so that you can realize success and become motivated to achieve greater goals. As you begin to experience success, your self-confidence will improve and inspire you to achieve even greater results.

INTRODUCTION TO TRAINING DESIGN

The three key ingredients of successful adult training are *presentation, application,* and *feedback*. Presentation is the delivery of content to the learners in a training session. Traditionally, this method has been delivered in a lecture format. Because adults learn in different ways from children, additional instructional methods should be utilized. Demonstrations and working in groups are some of the presentation methods that have proved to be more successful with the adult learner.

The application phase of training begins when learners are given an opportunity to practice the knowledge and skills they have developed in the presentation phase of training. Often the learner is not presented with this opportunity until he is on the job. Knowing that adults want to take an active role in their learning, the benefit of application during training is obvious. Role-playing realistic scenarios and practice exercises are some of the ways in which adult learners can practice the skills needed on the job.

For maximum learning effectiveness, the feedback phase of training should occur directly in conjunction with the application phase. When a learner is attempting to practice knowledge and skills or to apply concepts learned in training, he or she should be given immediate feedback on performance. The true test of whether an adult has learned a new skill or not is whether the learner can actually apply what has been learned. Through a combination of application and feedback, trainers can get a much better grasp of whether this is occurring.

CONCLUSION

To educate is to train someone for a skill, trade, or profession by formal instruction and supervised practice. An effective driver trainer will strive to advance and develop the standards of instruction provided in the classroom. The successful driver trainer will establish a role model of professionalism for his students in every capacity, including the need to commit to lifelong learning in order to maintain the high standards necessary for this career field.

Suggested Ideas for Trainer Training Exercises

1. Why did you decide to become a driver trainer?

2. What are your professional goals for one year and five years into your career as a driver trainer?

3. List the specific assets you feel you bring to the job as a driver trainer.

4. Identify the key characteristics of an effective driver trainer. Explain how these factors contribute to successful teaching.

5. List the three key elements of training design.

Teaching the Adult Learner

OVERVIEW

Adult learners present unique challenges for driver trainers because of the way they approach a learning situation. For trainers of adult learners to be effective, it is important to understand these students. Adults are not passive learners; they want to be able to immediately put to use new information they acquire. They bring a wealth of experience to the learning process, as well as the expectation that the content and instruction of learning will meet their specific needs for information. This chapter will help explain the unique characteristics that adult learners bring to the training situation and suggest strategies to use for effective training. Adult learner characteristics should become guidelines as driver trainers approach each day of teaching in the classroom, on the range, and on the road.

The materials in this handbook are designed for the person who is creating the trainer training program. These materials offer suggestions for objectives for working with trainers of adults, as well as suggested exercises, from which the trainer may select items to use in a learning situation.

OBJECTIVES

As a result of reading and studying this chapter, you should be able to:

- Identify the unique characteristics of adult learners.

- Understand how adult learners receive and process information.

- Understand how individual differences influence training for adults.

- Identify four major ways in which people learn.

- Identify strategies that address the different ways in which adults learn.

CHARACTERISTICS OF ADULT LEARNERS

Driver trainers should consider the following key points identified by experts concerning how adults learn. Understanding your learners is the first step in designing effective training situations.

The following points highlight important characteristics of adult learners:

❑ Adult learners are goal oriented.

❑ Adult learners are self-directed.

❑ Adult learners approach new learning based on their past learning experience.

❑ Adult learners want to be treated with respect.

❑ Adult learners bring specific habits and opinions to the learning process.

❑ Adult learners relate new information to prior knowledge.

❑ Adult learners expect to be active participants in learning.

Before a driver trainer begins an educational program, it is important to have a solid understanding of who the learners will be. It is important to know why they are being trained, what experience they bring with them to the classroom, and any anxieties and expectations they have. By understanding your learners, you will take the first important step in becoming an effective driver trainer.

A number of characteristics have been identified by researchers as common to all adult learners.

Goal oriented Adult learners are goal oriented and bring specific expectations to the learning situation. That is, they want information they can understand and put into practical use right away. Adults need immediate feedback so they can know whether or not they understood the information that was just given them. If trainees can obtain this type of information during the first training session, they will get the feeling of being able to accomplish required tasks, which will build a foundation for successful learning. Effective trainers stick to teaching the information they have been hired to present, focusing on the purpose of the learning experience. In order to respond to these factors, effective trainers should give adult learners an outline of what they will teach and what the students must do. This will enable students to measure where they are along the continuum of learning. For example, students should be able to discern *where they are* (at the beginning of the training);

where they are going (what will result from their learning efforts), and *how they are going to get there.*

Self-directed and autonomous   Adults want to direct their own learning. Teachers of adult learners must actively involve adult participants in the learning process, including getting their perspectives on the topics to be covered and letting them work on projects that reflect their individual interests. Participants should be expected and allowed to take responsibility for their own learning, including setting learning goals, locating appropriate resources, deciding on the types of learning methods used, and evaluating programs. Building lesson plans around learners' needs and guiding participants to their own knowledge rather than supplying them with facts will help the learners understand how the class will help them reach their goals, as well as help to build learner capacity to use new information.

Past learning experience   Adult learners typically have experienced a wide variety of training in the past, both positive and negative. If adult learners did not do well in school, for example, they may be apprehensive or defensive when placed in an adult-learning environment. First-time trainers often make the mistake of treating adult learners like students in a traditional school setting. This type of treatment can also cause learners to be defensive. It is important for trainers of adult learners to consider the past experience of their learners in designing and delivering training sessions.

Need for respect   Adult learners need to feel respected for what they know and need to be treated as adults. Adults learn best in a safe and supportive environment. They should be treated as equals in experience and knowledge and allowed to voice their opinions freely in class. Trainers of adult learners report that they often learn as much from their students as their students learn from them. The learning environment created by the trainer should encourage intellectual freedom and experimentation. Adult learners do not like to make mistakes in public and do not appreciate being embarrassed. If a trainer of adult learners must correct a problem or offer criticism or suggestions, this feedback should be provided in a correcting and coaching manner that demonstrates respect for the individual trainee.

Established habits and opinions   Adults learners bring set behaviors to the learning experience that may make them resistant to learning if they feel these behaviors are being challenged. The more extensive and varied a person's experience, the easier it is for that person to acquire more knowledge. Adult learners also bring established opinions to the learning process. These opinions may get in the way of the learning progress, but adult learners need to feel like their ideas and opinions have value and weight. Whenever possible, the trainer should take advantage of the learner's experience to improve training techniques and procedures.

Relationship to prior knowledge   Adult learners tend to make connections with information or knowledge they already have. Effective trainers of adult learners relate new information to situations or procedures with which trainees are familiar in order to make acquiring new knowledge less threatening. New information that conflicts with what is already held to be true and forces a reevaluation of the old material is integrated more slowly. Instruction should move from the known to the unknown in a logical and measured progression. This technique makes the gradual acquisition of new knowledge more practical, useful, and successful.

Active participants in learning   Adult learners need to participate in the learning process. An effective technique for trainers of adult learners is to limit lecture time to essential information and then engage trainees by asking questions about reactions and opinions. Adult learners should be challenged about how they will use new information. If learners have an active role in the learning process, they will have a greater stake in making that process successful.

As an effective trainer, it is important to be aware of learners' attitudes, past experiences, habits, and opinions in order to understand their perspectives. In doing so, you will be able to help them discover how useful a change in behavior can be and engage them in the learning process to enable changes to take place more quickly.

To keep pace with the rapidly changing world we all face, learners today must be equipped for lifelong learning. An effective trainer can help equip students for the demands of today's workplace through encouragement and support of the learning process. In addition, today's students are more diverse than ever. For example, women and minority groups that have not previously been present in large numbers in truck driving are now entering the field. The driver trainer can help to supply trucking companies with these new sources of labor, as well as helping new entrants into the field equip themselves for new experiences. It is important for trainers of adult learners to understand the varied backgrounds of their students in order to better prepare the learning environment and to ensure student success.

Success in learning stimulates the desire for additional learning. By presenting information in manageable doses, the trainer enables the learner to experience success. With time, experience, and growth on the part of trainees, you can increase the size of the information segments, allowing for your learners to be challenged while they experience learning success.

HOW ADULTS LEARN DIFFERENTLY

Good trainers realize that individual trainees differ in mental ability as much as their physical makeup and make provisions for adapting their techniques to fit the learning abilities of each individual.

According to learning theorists, there are four major ways in which people learn. Schools traditionally teach in only one of these styles; therefore, if your learners are not the kind of students who like lecture, reading, and writing, school will be difficult for them. Adult learners utilize a preferred style for learning that defines who they are.

In this chapter, driver trainers will learn how to identify the various learning styles of adult learners and practice a variety of teaching methods that will facilitate learning for all students, regardless of their learning style. It is essential to learn tools and techniques that create energy, enthusiasm, interest, and variety. Techniques for handling special learning needs and helping students overcome barriers to learning are also needed. This chapter will introduce a successful way to teach all learners.

THE 4MAT© SYSTEM: A CYCLE OF LEARNING

4MAT is a natural cycle of learning that all learners go through when they learn. It describes how different people learn according to different styles. 4MAT contains all the elements of learning driver trainers need to understand and practice in order to adjust their approach according to learner needs. When educators understand this process and use the cycle to teach, they improve the learning of their students.

The following information about the 4MAT system is adapted from materials developed by the system's creator, Bernice McCarthy, PhD. The use of the 4MAT materials will assist driver trainers to understand and teach to their students' diverse learning styles.

Different people perceive and process experience in different preferred ways. These preferences dictate learning styles. It is important that learners understand their preferences and are comfortable with the way in which they learn the best. Learners have a preference for one mode, but this does not mean they cannot function effectively in other modes. In fact, learners who can move from one mode to another according to the requirements of the learning situation are at a distinct advantage over those who limit themselves to the mode in which they are most comfortable.

The underlying theory of the 4MAT learning cycle was developed based on observing how children learn. Think about how a child learns. When a child learns, he is absorbed by what is and naturally reacts to what he experiences. He is unencumbered by fear of failure as he approaches the learning process. The child is and the child does, thus the pattern of being someone and reacting to the world develops as he embraces the learning cycle and the purpose of learning—to find meaning in the world. Humans need to learn how to be someone as well as how to react to the world; in other words, humans need to learn how to learn.

Perceiving is how humans take in experience. *Processing* is how they react to, confront, and resolve what happens to them. The two activities, *perceiving* and *processing*, are critical learning elements. Understanding how each works is the core of how learning happens.

To understand the 4MAT learning style system, picture the learning cycle like a clock with learning starting at 12 o'clock and continuing around the clock until the next 12 o'clock (Figure 2-1). Visualize the first dimension, *perceiving*, as a line running from 12 o'clock to 6 o'clock. This line represents *perception*, or how people take in the things they learn. Visualize the second dimension as a line running from 3 o'clock to 9 o'clock. This line represents *processing*, or what learners do with the information they take in and reflect on. Learning grows out of the natural cycle of perceiving and processing.

Perceiving

Learners perceive in two ways. They *feel* their experiences (the 12 o'clock place), and then they *think* about their experiences (the 6 o'clock place). When adult learners encounter new information, first they grasp the experience and then they reflect on what it is they have encountered. Perceiving is represented by a line between 12 o'clock and 6 o'clock. In order to deal with new information, learners have to step back from what they have encountered and begin to analyze what they have experienced. Once learners encounter and analyze new information, they place it somewhere that connects it to past knowledge and understanding and order it for themselves so they can place it in the context of their own world.

Individual learners vary in how they utilize perceiving and processing. Those who judge by feeling first do not really understand those who

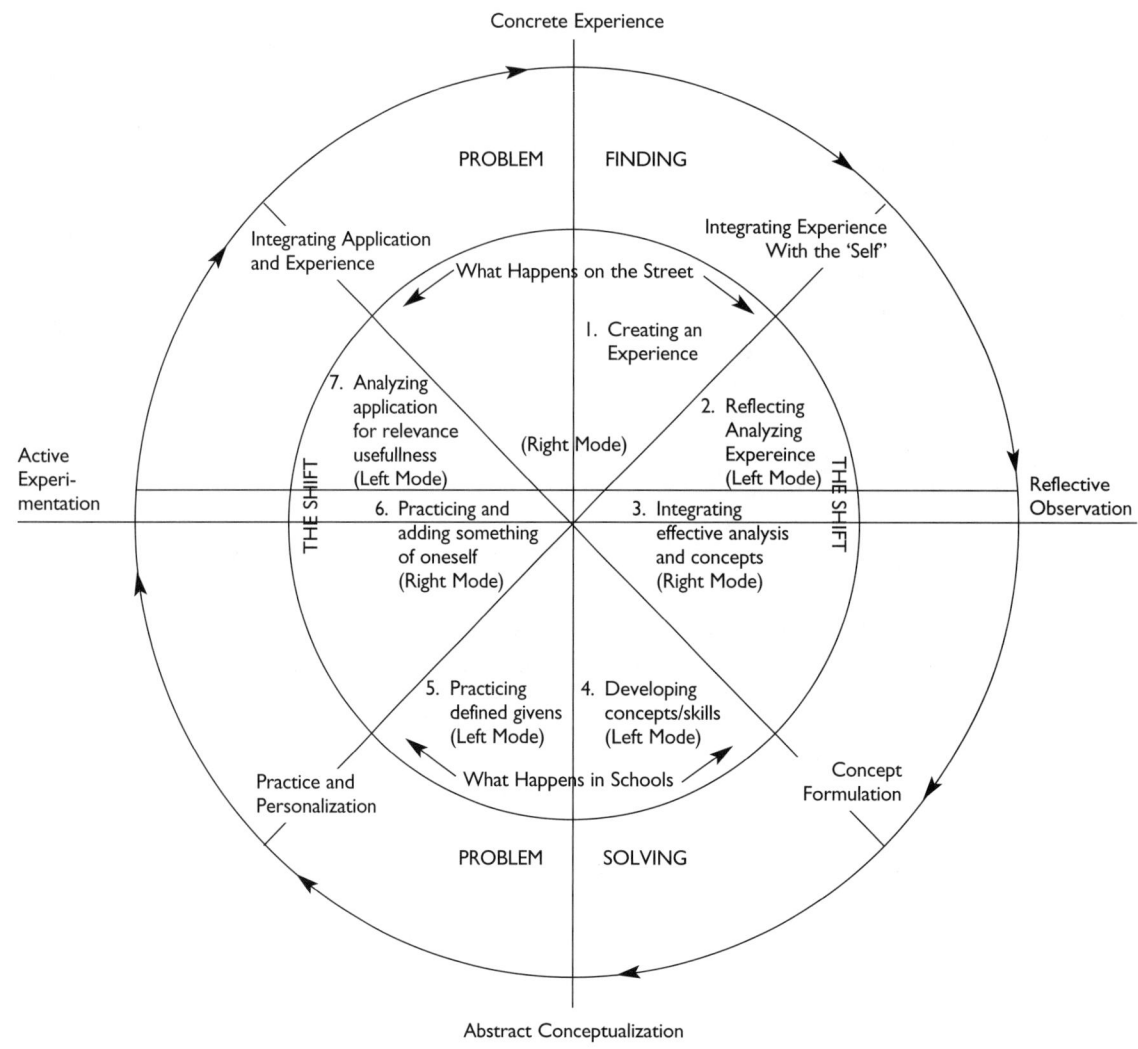

FIGURE 2-1   The Complete 4MAT System Model

approach new things by thinking first. Those who reason first do not really understand those who feel first. We each develop ways to perceive and process information based on our instincts for using approaches with which we are more comfortable. Some learners favor and trust the feeling way of perceiving more than the thinking way. Others trust the thinking way more than the feeling way. This variety in approach makes a great difference in how people learn.

Both ways of perceiving are valuable and necessary. After encountering an experience, learners need to order their experiences, to make sense of them, to name and classify them. The tension between these two ways of perceiving is the central dynamic in learning. Real learning is achieved only through a balance between the two and an understanding of the utility of both approaches. For example, those who favor the feeling dimension need to understand the benefit of the thinking dimension. Those who move to thinking too quickly should linger in the feeling of things: perceptions and concepts, experiencing and conceptualizing. The most important element in all this is the growth of learning itself. Unless learners move out of experience, or the 12 o'clock place, into ordering their experiences and understanding what happens to them by classifying and naming, they will never learn. It is the act of moving from feelings to thoughts that gets learners from being in their perceptions to being able to see the perceptions themselves. By moving from the 12 o'clock feeling place to the 6 o'clock thinking place, individuals can focus on how they are separate from their perceptions and how they can differentiate themselves from them.

To better understand this, think about how you experience anger. When you feel angry you are *in* the anger, or in the 12 o'clock place. If you suddenly stop and ask yourself, "What am I so angry about?" you make the anger you are feeling an object of thought. That is the 6 o'clock place. If a person can stand outside of the anger to know the inside of it better, he can learn from it. All learning demands that learners step outside of experience in order to understand. Conceptual thought demands separation of thinking from feeling, of object from subject.

The challenge in learning is how to balance one's feelings with one's thoughts and how to move around the cycle continuously. Human beings experience all newness by connecting it to oldness. As people begin to learn, they place the learning into the areas they already know; they name the new learning, which makes it more accessible to them. Think of how you are able to search for a specific file in a filing cabinet or in your computer because you know the name of the folder.

But the name for something is never the thing itself, although some people think so. Teachers often ask students to memorize the names of things or a sequence of how to do things without allowing them to try things for themselves. No reality can ever be adequately described in words, as words are just the derivative sign language we use to describe things. They are not the same as real things. Educators must encourage their students to learn new things by using thoughts and feelings so they can develop a deeper and richer perception, becoming more informed because they have learned about themselves in the process.

Processing

In addition to perceiving, or taking things in, people must do something with the things that they are experiencing if they are to truly learn. This is how people process what they learn. They deal with their learning in some way that helps them to use and integrate it so that it becomes a part of their lives. As with perception, humans process in two ways: They reflect on their experience and they act on that reflection; they watch and they do. If what they encounter is not used, it is not learned. On the processing continuum in the previous figure, *acting* is at the 9 o'clock position and *reflecting* at 3 o'clock.

If a person is to create meaning from his reflections, he needs to act on those reflections. People must do something with the results of what they watch—to make the results useful or to extend those reflections into testing them. As with feeling and thinking, reflecting and doing should be balanced. When people reflect on their actions, they see how they do not follow through. These kinds of reflections give learners purpose. When people act with purpose, they give meaning to their actions.

The Complete Learning Cycle

To combine perceiving (taking things in) with processing (doing something with them) creates a natural learning cycle. Relating perceiving and processing, or overlaying the 3 o'clock to 9 o'clock dimension on the 12 o'clock to 6 o'clock dimension, reveals a cycle that contains the essential elements of learning: feeling, reflecting, thinking, and doing.

People start with their experience at 12 o'clock, with how they "feel about the world." Next, they reflect. Not only do they watch what is happening, but they also examine how they feel about what is happening. This is the 3 o'clock place. Then they conceptualize what is happening or they develop an idea about it—the 6 o'clock place. Finally, they act on it or make it their own; then it is learned and integrated into their lives. This is the 9 o'clock place. That is the complete learning cycle. Learning is not complete until the learner makes what he has learned real in his life. The student must connect and integrate what he has learned.

Learning begins with individuals and their connections to their experience, and it ends when they adapt their use of the experience in their lives. The learners are dealing with themselves as well as listening to others in this process. They have reflected and listened—first to their own voice blending their experiences with the voices of others—before integrating new information into their lives. The main purpose of the 4MAT learning cycle is to help educators and trainers bring balance to learning, so that students will come to speak about what they have learned in their own voices.

If you follow the clock analogy, you realize that you can only get from 12 o'clock to 6 o'clock by going through 3 o'clock, and you can get back to 12 o'clock from 6 o'clock only by going through 9 o'clock. In other words, individuals start from a place of personal knowing in they way they see and feel things. They rely on their senses and begin to trust their intuitions grounded in their experiences. They build on the known as they move to the unknown, to deeper beliefs, to changes because the cycle spirals. It begins with feeling, with heart, with intuition, with emotion, with a person's reality.

Reflections move people to concepts—first they image them—that is, they see pictures of them in their minds; then they name them and develop theories about them. Think of this as *conceptualizing* or 6 o'clock. Next people apply their theories to test them or see things in light of their theories. Think of this as *doing* as 9 o'clock. If something works, if it makes useful sense, people integrate it into their lives, or back to the 12 o'clock position.

The patterns and combinations that people choose create their learning preferences. Learners begin with feelings, bringing all of who they are to the act, beginning the process of internalizing what has happened, pondering, musing, separating themselves from their feelings in order to look at their feelings. They move to naming and conceptualizing so as to understand, to be objective, to organize what they encounter before they begin to integrate it.

Learning Style Characteristics

Each person travels the cycle according to parts of it that feel more natural. Some linger in the 12 o'clock feeling place, where they process things filled with the moment and are open to sensations. Some stay in the 3 o'clock reflecting place, where they ponder how they feel. Some tend toward the 6 o'clock thinking place, where they stand back and examine and name it. Some linger in the 9 o'clock doing place, where they explore, act, and create outcomes. The comfort that individuals feel at certain places along this learning cycle creates patterns. These patterns influence how they approach newness, how they deal with ambiguity, the things they tend to avoid, the speed with which they move to judgment, the depth of their reflections, and the boldness of their actions.

The intersection of the feeling 12 o'clock—thinking 6 o'clock line with the acting 9 o'clock—reflecting 3 o'clock line creates four quadrants (see Figure 2-1). People who linger in feeling and watching have a particular learning style. These people are in the first quadrant, the 12 o'clock to 3 o'clock place, bounded by feeling at 12 o'clock and reflecting at 3 o'clock. They are called *type 1, innovative learners*, because they feel and reflect on their experiences and are primarily interested in personal meaning. They typically need reasons for learning that connect new information with personal experience and establish that information's usefulness in their daily life. Cooperative learning, brainstorming, and integration of content areas are some of the types of instruction that work best with these types learners.

People who are comfortable watching and thinking also have a particular style. These people are in the second quadrant, the 3 o'clock to 6 o'clock place, bounded by reflecting and thinking. These are *type 2, analytic learners*. They are primarily interested in new facts in order to deepen their understanding. They reflect on their experiences, analyze them, and think about them. Direct lectures and independent research are instruction modes that work well with these learners. They enjoy analyzing data and researching expert opinions.

People who are comfortable in thinking and doing have another style. These people are in the third quadrant, the 6 o'clock to 9 o'clock place, bounded by thinking and doing. They are *type 3, common sense learners*, because they think about theory and act to apply it. They are primarily interested in how things work and want to roll up their sleeves and try out new learning. Concrete, experiential learning activities work best with this group, such as hands-on tasks.

People who like to linger in doing and feeling have another learning style. These people are in the fourth quadrant, the 9 o'clock to 12 o'clock place, bounded by doing and feeling. They are *type 4, dynamic learners*, because they embrace their experiences and act to extend and enrich them. These learners tend to rely heavily on their own intuition and seek to teach themselves and others. Role-play and simulation activities are instruction modes that work well with these learners.

The following information outlines implications for how these different types of learners function in various roles and how preferences for taking and processing new information reflect different characteristics. This information has implications for structuring learning to accommodate different learning styles.

CHARACTERISTICS OF LEARNING STYLES

Type 1: The Innovative Learners

Innovative learners perceive information concretely (experiencing it) and process it reflectively (pondering it). They:

- ❏ Integrate experience with self.
- ❏ Learn by listening and sharing ideas.
- ❏ Value insightful thinking.
- ❏ View their experiences from many perspectives.
- ❏ Love harmony and work diligently to bring it to the lives of the people around them, both personally and professionally.
- ❏ Are committed to whatever task they undertake and seek commitment in others.
- ❏ Are fascinated with people and cultures.
- ❏ Absorb reality.
- ❏ Need clarity in their lives.

Meaning is first and foremost what they seek as learners. Sometimes their feelings interfere with their common sense. As *educators,* they are primarily interested in helping their students achieve personal growth. Self-awareness is a major goal that they have for their students. They:

- ❏ Organize curricula whenever possible to help learners become more authentic human beings.
- ❏ See knowledge as assisting growth toward self-actualization.
- ❏ Organize group work and discussions and encourage honest feedback about feelings.
- ❏ Organize cooperative studying.
- ❏ Focus on meaningful goals.

As *leaders,* they take time to help employees develop good ideas. They:

- ❏ Tackle problems by reflecting alone and then brainstorming with staff.
- ❏ Lead with their hearts and engage others in decision making.
- ❏ Exercise authority through trust and participation.

❑ Work for organizational solidarity, in dialogue with staff concerning the mission.

❑ Need staff who are supportive and share their sense of vision.

As *parents,* they try to facilitate self-actualization with their children. They:

❑ Help their children become more self-aware.

❑ Believe learning should enhance the ability to know oneself and one's place in the world.

❑ See knowledge as enhancing personal meaning and relationships.

❑ Actively encourage speaking about feelings.

❑ Conduct family discussions with honest and realistic feedback.

❑ Try to engage family members in cooperative efforts.

❑ Help their children identify meaningful goals.

❑ See discipline as necessary to help their children understand life.

❑ Tend to worry excessively and sometimes are too easygoing.

Strength:	Nurturing spirit and imaginative ideas.
Function by:	Clarifying values.
Goals:	To be engaged in important issues and to bring harmony.
Favorite question:	Why?

Type 2: The Analytic Learners

Analytic learners perceive information abstractly (thinking it) and process it reflectively. They:

❑ Form theories and concepts by integrating their observations with what they know.

❑ Seek coherence and continuity.

❑ Strive to know what the experts think.

❑ Learn by carefully reflecting on and thinking through what they are learning.

❑ Are great detail people.

❑ Work sequentially.

❑ Critique data and information.

❑ Are thorough and industrious and will reexamine facts assiduously until they understand.

❑ Try to maximize certainty and are uncomfortable with subjectivity.

Schools are perfectly designed for the way analytic learners learn. Sometimes these learners cannot see the forest for the trees. As *educators,* they are primarily interested in transmitting knowledge. They:

❑ Strive to know their content well, to be scholarly.

❑ Believe curricula should further understanding of significant information, should be presented systematically, and should deepen comprehension.

❑ Encourage outstanding learners.

❑ Like their students to present detailed facts precisely and sequentially.

❑ Believe in the rational use of authority.

As *leaders,* they assemble facts and data into coherent theories. They:

❑ Tackle problems with logic.

❑ Lead by principles and procedures.

❑ Exercise authority with assertive persuasion, knowing the facts.

❑ Work to enhance their organization as an embodiment of tradition and prestige.

❑ Need staff who are well organized, have things down on paper, and follow through on decisions quickly and precisely.

As *parents,* they want their children to have important knowledge. They:

❑ Try to help their children be as accurate as possible.

❑ Believe learning should deepen the understanding of significant information and be presented systematically.

❑ Want their children to understand how the world works.

❑ Encourage their children to be outstanding lifelong learners.

❑ Provide a structured and organized home environment.

❑ Seek to help their children love knowledge.

❑ Believe diligence and organization are necessary for success.

❑ See discipline as necessary to enable their children to understand the kind of behavior that society expects.

❑ Tend to be rigid and sometimes discourage original, creative thinking.

Strength: Creating concepts and models.

Function by: Thinking things through.

Goals: Intellectual recognition.

Favorite question: What?

Type 3: The Common Sense Learners

Common sense learners perceive abstractly (thinking things) and process actively (by doing things). They:

❑ Integrate theory and application.

❑ Learn by testing theories and applying common sense to them.

❑ Are pragmatic; if something works, they use it.

❑ Are down-to-earth problem solvers who resent being given answers.

❑ Do not stand on ceremony but get right to the point.

❑ Have a limited tolerance for fuzzy ideas.

❑ Value strategic thinking.

❑ Are skills oriented.

❑ Like to experiment and tinker with ideas and things.

❑ Instinctively understand how things work.

❑ Edit reality, cut right to the heart of things.

They sometimes seem bossy and impersonal. As *educators,* they are primarily interested in productivity and competence. They:

❑ Strive to give their students the skills they will need in life.

❑ Believe curricula should be geared to competencies and economic usefulness.

❑ See knowledge as enabling learners to be capable of earning their own way.

❑ Encourage the practical application of learning.

❑ Like hands-on activities and teach skills well.

❑ Believe the best way is determined scientifically.

❑ Use measured rewards.

As *leaders,* they thrive on plans and time lines. They:

❏ Tackle problems by making unilateral decisions.

❏ Lead by personal forcefulness and often inspire real quality.

❏ Exercise authority with reward and punishment.

❏ Work hard to make their organization productive and solvent.

❏ Need staff who are task oriented and move quickly.

As *parents,* they foster competence. They:

❏ Try to give their children the skills they will need in life.

❏ Believe learning should improve job skills and real-life quality.

❏ View knowledge as enabling their children to find their own way.

❏ Encourage their children to find practical applications for what they learn.

❏ Like active, hands-on family projects.

❏ Believe success is best judged by whether something works.

❏ Use measured rewards.

❏ See discipline as necessary to enable children to eventually take their own power.

❏ Tend to be inflexible and sometimes lack the ability to express feelings.

Strength:	Practical application of ideas.
Function by:	Gathering factual data from kinesthetic, hands-on experience.
Goals:	To align their view of the present with future security.
Favorite question:	How does this work?

Type 4: The Dynamic Learners

Dynamic learners perceive information concretely (experiencing it) and process it actively. They:

❏ Integrate experience and application.

❏ Learn by trial and error.

❏ Believe in self-discovery.

❏ Are very enthusiastic about newness.

❑ Are adaptable, even relish change.

❑ Excel when flexibility is needed.

❑ Often reach accurate conclusions in the absence of logical justification.

❑ Are risk takers.

❑ Are at ease with people.

❑ Enrich reality by taking what is and adding to it.

Sometimes dynamic learners are manipulative and pushy. As *educators,* they are primarily interested in enabling learner self-discovery. They:

❑ Try to help their students act on their visions.

❑ Believe curricula should be geared to learners' interests.

❑ See knowledge as necessary for improving the larger society.

❑ Engage in and encourage experiential learning.

❑ Use a variety of instructional methods.

❑ Are dramatic and seek to energize their students.

❑ Attempt to create new forms, stimulate new life, draw new boundaries.

❑ Tend to be rash and manipulative.

As *leaders,* they thrive on crisis and challenge. They:

❑ Tackle problems by looking for patterns and scanning possibilities.

❑ Lead by energizing people.

❑ Exercise authority by holding up visions of what might be.

❑ Work hard to enhance their organization's reputation as a frontrunner.

❑ Need staff who can follow up and implement plans and details.

As *parents,* they foster self-discovery. They:

❑ Try to help their children have visions and act on them.

❑ Believe learning should be geared to their children's interests.

❑ View knowledge as necessary for improving the larger society.

❑ Encourage experiential learning for their children.

❑ Like humorous and challenging (often competitive) family activities.

❑ Are dramatic and entertaining.

❑ Help their children seek new boundaries.

❑ See discipline as necessary to enable their children to become self-disciplined.

❑ Tend to take on multiple activities, sometimes leading to inconsistencies and punishments that do not fit the crime.

Strength:	Action, getting things done.
Function by:	Acting and testing experience.
Goals:	To bring action to ideas.
Favorite question:	What if?

All four styles of learning are equally valuable. Each has its own strengths and weaknesses. Regardless of your type and preferences, your style is how you learn.

Challenges for the Four Styles

Each learning type presents specific challenges for the learner. The challenges for the type 1, or innovative learner, are to act more quickly and to get the job done, while understanding that people need time to process information at their own pace. The challenge for the type 2, or analytic learner, is to remain open to ambiguity and the unknown while understanding the need for precision and the right data. The challenge for type 3, or common sense learners, is to listen and understand the processing time others need to discover things for themselves while acknowledging the need to act and get the job done. For the type 4, or dynamic learner, this challenge is to develop structure while covering a lot of territory in order to break boundaries.

Whatever your least-developed quadrant is, that is the source of the most energy, the area in which you can push boundaries in order to experience real growth. Growth comes from moving around this cycle. In the learning process, people start with their perceptions, then separate themselves from their perceptions by beginning to look at them as separate objects, acting on their understanding, refining it, and editing it, so they can come full circle or complete the cycle with a new understanding.

It is reasonable that people operate from their perceived strengths, but real growth requires the entire cycle. As learners move through life, many of them come to conscious awareness of the entire cycle. They search for meaning, recognize the importance of the big picture, and attempt to apply what they have learned to their daily lives. Learners would be better equipped to find balance in their lives and work if they were introduced to the cycle and understood the strengths inherent in each approach.

BALANCED LEARNING AND THE 4MAT CYCLE

The 4MAT cycle is a method for instructors to help their students become authors of their own learning process. The original 4MAT model was based on the belief that all learners needed to master all the quadrants. The use of the 4MAT cycle during the last twenty years has proved that if learners function well in all parts of the cycle, wholeness and learning balance is the result. The flow from personal meaning in quadrant 1 back to the integration of learning in quadrant 4 allows and encourages individuals to bring their own adaptations to what they are learning. The structure inherent in the expert knowledge in quadrant 2 and the practice of that knowing in quadrant 3 give the learner the expertise and confidence to move into unique, personal adaptations. Real learning moves from the personal connections of quadrant 1, to the conceptual knowing of quadrant 2, to the practice and tinkering of quadrant 3, and then to the creative integration of quadrant 4. The 4MAT cycle allows this movement to happen as learners travel the cycle according their own preferred learning style.

Knowledge must be used, and because all humans are unique, they integrate what they learn in their own ways. What they learn is transformed into a particular use, a distinct way of doing. It is in the transformation that real understanding happens.

4MAT is formed from the rhythm of the perceiving and processing dimensions of the natural learning cycle. These four quadrants embody the core elements of learning: feeling, reflecting, thinking, and doing. The first quadrant combines feeling and reflecting, the heart of meaning. The second quadrant combines reflecting and thinking, the heart of conceptualization. The third quadrant combines thinking and doing, the heart of problem solving, and the fourth quadrant combines doing and feeling, the heart of transformation.

Each of the core elements of learning—feeling, reflecting, thinking, and doing—elicits a different question from the learner. Successful learning answers four questions: Why? What? How? and What if? Quadrant 1 creates personal meaning. Learners question the value of new learning by connecting it to themselves. The question to be answered is "Why is this of value personally?" Quadrant 2 creates the conceptualized content, structuring knowledge into units that form the essence of new ideas. The question to be answered is "What is out there to be known?" Quadrant 3 creates usefulness, the transferability into life, problem-solving skills. The question to be answered is "How? How does this work?" Quadrant 4 creates integration. This is where the learner adapts the learning into something new and unique for him, in his own life. The question to be answered is "What if? If something is used in this way, what will happen?"

The cycle brings balance and wholeness to learning and is the way in which people make learning meaningful. As the learner's style is formed by moving through the cycle, he moves among his comfort zones to the edge of his competence. With a supportive learning environment the learner can more confidently move among the learning places where growth takes place.

LEARNING AND THE BRAIN

The two hemispheres of the human brain process information in different ways. Learners tend to process information using whichever hemisphere is dominant, the left side or right side of the brain. Learning is enhanced when people stretch themselves and use their less dominant side, like using the learning style quadrants with which they are less familiar.

Current studies regarding right-brain and left-brain processing help to support the understanding of learning cycles and learning styles. Trainers seeking to understand how students approach learning should consider the significance of left-brain and right-brain studies. The two sides of the human cerebral cortex function differently, as they support two different mental operations with separate modes of consciousness. Normally, the two sides work in cooperation.

The left mode of the brain:

❑ Operates with analysis and examines cause and effect.

❑ Breaks things down into parts and examines or categorizes.

❑ Seeks and uses language and symbols.

❑ Abstracts experience for comprehension, generates theory, and creates models.

❑ Is sequential and works in time.

The left mode is linear. Its thinking skills are of high order and use language, analysis, sequence, logic, and objectivity. Linear thinkers complete a task with time left over, produce specific and clear instructions, stay on task, hear exactly what is said, and think before acting. Their speech is linear, one word or thought following another in a structured, sequential manner. Linear thinkers are frustrated by ambiguity; they are not at ease when they cannot specify.

The right mode of the brain:

❑ Operates intuitively.

❑ Sees wholes, forms images, and creates mental combinations.

❑ Seeks and uses patterns, relationships, and connections.

❑ Manipulates form, distance, and space.

❑ Is simultaneous.

The right mode is nonlinear or round. It knows more than it can tell, filling in the gaps, clearing space, and seeing the partially eclipsed. It creates connections and solves problems sideways, circling, flanking, then all at once. The right brain is a great storyteller, a consummate musical moment, a great performance, a perfect sentence, knowing instantly that something is true, understanding how all the parts fit, doing more than one thing at a time. The right mode functions without words, using pictures, creating metaphors, picking up tones and emotions.

If educators continue to focus on only one mode of processing, they do great harm to the whole brain. If they incorporate round, nonverbal thinking into learning and working fully with left-mode tasks, more of their students will have an opportunity to succeed.

The instructional sequence suggested by Bernice McCarthy in the 4MAT learning system teaches to the four styles using both left- and right-brain processing techniques. Integrating the four learning styles and processing modes ensures that trainers are meeting all learners' needs.

The 4MAT system is designed to provide every student with a preferred task during every lesson. Figure 2-2 lists the eight instructional events proposed by this system.

QUADRANT LEARNING GOALS

Each of the four quadrants lends itself to different goals. The goal in quadrant 1 is personal connections—meaning brought from the past into discussions. The learning climate needs to be easy, open, and nurturing. Use experiences, storytelling, dialogue, and listening that is interested

Step	Left Mode	Right Mode
	Why? (Motivate and Develop Meaning) I	
I		Create an experience (CONNECT)
2	Analyze/reflect about the experience (Examine)	
	What? (Reflection and Concept Development)	
3		Integrate reflective analysis into concepts (Image)
4	Develop concepts/skills (Define)	
	How? (Usefulness and Skill Development)	
5	Practice defined "givens" (By)	
6		Practice and add something of oneself (Extend)
	What If? (Adaptations)	
7	Analyze application for relevance (Refine)	
8		Do it and apply to more complex experience (Integrate)

FIGURE 2-2   Integration of Learning Styles and Processing Modes

and focused. The method is discussion, initiating cooperative learning and trust. The educator is the initiator, the motivator capturing enthusiasm. The learner is receiving.

The goal in quadrant 2 is defining the learning—the best information, pertinent facts that are structured and planned. The learning climate is receiving, taking in, and being present to the content. The method is traditional teaching, including lecture and demonstration. The educator delivers the content, conceptualizing and theorizing. The learner is forming concepts. The goal in quadrant 3 is problem solving—practicing, recording, and placing learning into one's life. The learning climate is active, finding personal uses for the learning. The method is coaching, or using mastery learning to reach conclusions. The educator is coach, facilitating and setting up environments that encourage experimentation. The learner is tinkering, testing, trying. The goal in quadrant 4 is refining and creating—skills perfected and learning placed into one's life. The learning climate is dynamic and adapting learning into something unique in one's life. The method is self-discovery using learning independently and making better connections. The educator is cheerleader. The learner is adapting, acting, and creating.

In the process of going through the 4MAT cycle, from quadrant to quadrant, we slow down in our comfort places and stretch in those that challenge us. We need to make multiple options available to students. The educator is more active in quadrants 1 and 2, motivating, creating connections to students, and delivering content. In quadrants 3 and 4, learners take over the spotlight, becoming the center of activity. If this shift does not occur, the learning never moves from the educator to the learners. 4MAT clarifies how the criteria for learning changes as a person moves from place to place, from meaning to connection in quadrant 1, to concepts in quadrant 2, to skills and usefulness in quadrant 3, and, finally, to creative adaptation and modification in quadrant 4.

BIBLIOGRAPHY

Barnes, Letha. *Milady's Master Educator Student Course Book.* Albany, NY: Thomson Learning, 2001.

McCarthy, Bernice. *About Learning.* Barrington, IL: Excel, 1996.

McCarthy, Bernice. *The 4MAT System: Teaching to Learning Styles with Right/Left Mode Techniques.* Wauconda, IL: About Learning, 1981, 1987.

CONCLUSION

Adult learners have specific learning needs. Good trainers realize that individual trainees differ in their mental ability as well as in their physical makeup and make provisions for adapting their techniques to fit the learning abilities of each individual.

The 4MAT cycle is an instructional framework that addresses the four types of learning styles and is a method for instructors to help their students become authors of their own learning process. Different individuals perceive and process experience in different preferred ways. These preferences dictate learning styles. It is important that learners understand their preferences and are comfortable with the way in which they learn the best.

Suggested Ideas for Trainer Training Exercises

1. Think about past classes or training sessions you have completed as an adult learner and choose and discuss one that you feel was a positive experience. Relate your experience to the adult learner characteristic materials listed in the beginning of this unit and describe how these characteristics were handled. (The purpose of this exercise is to help you understand that successful training requires an understanding of these characteristics.)

2. List and discuss the differences between adult and child learners. Use the following list as a reference for review.

Adult	Child
• Voluntary learners	• Captive audience
• Problem-centered	• Subject-centered
• Experienced	• Inexperienced
• Learners decide content	• Teachers prescribe content
• Grouped according to interest or need	• Grouped according to age, level, and ability
• Concerned with using knowledge today	• Concerned with learning for the future
• View learning as a lifelong process	• View learning as terminal

	Adult	Child
	• On an equal relationship with teacher	• Subordinate to teacher
	• Flexible	• Rigid and traditional
	• Active learners	• Passive learners

3. Use the following list of basic concepts of adult learning. Ask trainees to form small groups to determine and list ideas under the "Implication" heading. Report results to the class and discuss ideas.

Concept	Implication
Self-directing	
Life experience	
Solve real-life problems	
Must want to learn	
Learn by doing	
Relate to previous learning	

4. Describe the two major elements of how people learn. In what two ways do learners perceive?

5. After experience, learners need to order their experiences, make sense of them, name and classify them, and anchor them in their consciousness. Describe how they do this.

6. As with perception, humans process in two ways. They _____ on their experience and they _____ on that reflection.

7. Perceiving (taking in things) combined with processing (doing something with them) creates a natural learning cycle. Explain how this relationship creates a natural learning cycle.

8. Each person travels the cycle of learning, but some parts of it feel more natural to a person than others do. Those are the parts in which learners tend to feel more comfortable. Why does lingering in these parts make such a great difference in the way people learn?

9. List at least five characteristics for each of the four types of learners: innovative, analytic, common sense, and dynamic.

10. List the four challenges for each of the four learning styles.

11. What is the favorite question in each of the four quadrants of learning?

12. List the goals identified for each learning quadrant.

13. Discuss the types of learning preferences of students you predict might be drawn to careers as truck drivers and why. Discuss sample learning exercises that might appeal to these types of learners.

CHAPTER 3

Classroom Management

OVERVIEW

The educator's ability to lead and inspire students to a sincere desire for learning is critical for successful outcomes. That ability must also be coupled with the educator's effective management of the classroom. For purposes of driver training, the classroom includes training conducted on the range and on the road. It should be the goal of every educator to create a positive environment in the classroom that will provide a pathway to career success for students. Effective classroom management requires that the driver trainer be prepared to understand learner behavior and develop methods for dealing with the various behavior encountered.

OBJECTIVES

As a result of reading and studying this chapter, you should be able to:

- Understand the principles of managing learner behavior.

- Understand how to create and maintain a positive classroom environment.

- Understand the impact of the type of image the trainer projects on learner success.

- Understand the basic principles of academic advising and counseling of students, as well as required administrative tasks.

CREATING A POSITIVE CLASSROOM ATMOSPHERE

Today's effective educators are essentially facilitators of learning for their students. To *facilitate* means to make easier or to bring about. Clearly, learning can be made easier for students if the classroom environment is well organized, pleasant, and conducive to learning.

The quality of learning and the potential for learning success is also affected by the physical environment in which it takes place. Driver trainers should strive to create a learning environment that is comfortable, free from distractions, safe, and conducive to group work. If possible, avoid setting up the room in the typical classroom style, with rows of seats facing the front. This type of physical arrangement might cue learners that they are simply going to be passive observers and they will be less likely to be active participants in the learning process. A classroom arrangement of round tables that can comfortably seat 4 to 6 people has been shown to be a more effective physical setting for adult learning. The trainer should move around the room during training and avoid standing behind a podium. The trainer should also always remain aware of comfort levels within the training room, including attending to such factors as lighting, air temperature, and reduction of any distractions.

PROFESSIONALISM IN THE CLASSROOM

The educator's image, attitude, and actions provide an important role model for students and can enhance their learning. Therefore, it is important that educators project the most professional image at all times. The educator must demonstrate a positive enthusiasm for the learning process. By establishing their own credibility, driver trainers can maintain order and control in the classroom, earning the respect of the learners and generating a high degree of cooperation in the learning environment.

MANAGING STUDENT BEHAVIOR

Setting well-defined goals, guidelines, and expectations for learner behavior early in the educator-learner relationship establishes a foundation for desired learning outcomes. For adult learners, it is important that driver trainers include their students in establishing these guidelines. The more active adult learners are in the learning process, the greater the probability of success.

Driver trainers need to provide role models of the types of behavior and professionalism that will be expected of their students on the job. Establishment of guidelines and rules of conduct will assist learners in understanding what is expected of them in terms of behavior and performance. Effective guidelines will help to prevent learner behaviors that might detract from the educator's ability to facilitate the learning process for all students. Prevention of undesired behaviors can be achieved through:

❑ The specification and maintenance of expected behavioral standards.

❑ Consistency in enforcing rules and applying consequences.

❑ Modeling appropriate behaviors.

❑ Continual monitoring of learner conduct.

Behavioral research indicates that learners will react more favorably and desired behaviors can be shaped more quickly and permanently through positive means rather than punitive actions. Educators can achieve more cooperation from learners by focusing on the positive. However, misconduct cannot be eliminated simply by establishing rules and guidelines; disciplinary procedures also must be established to effectively handle situations of noncompliance with expected behavior. Effective educators can model desired behavior and teach students to develop self-control—techniques that will serve the student well once she is in the stressful job of a truck driver.

Techniques used to address misconduct should be carefully selected on the basis of school policies, safety requirements, personality and personal style of the educator, nature and circumstances of the misconduct, availability of outside help if needed, and the time allowed to address the matter. The driver trainer must decide whether any disciplinary techniques should be applied immediately or later, such as at the end of the class or the end of the day.

Upon observing student misconduct, the driver trainer should alert the student that the misconduct has been observed and request that the behavior cease. If the behavior does not cease after an initial request, the driver trainer must decide on appropriate action to take based on her assessment of the situation as well as school policies. These policies should be clearly stated and readily available to all students.

Depending on the nature of the misconduct, the instructor might need to schedule a conference with the student to discuss the problem and potential solutions. The goal of any conference would be to help the learner understand specifically how the misconduct hinders learning in the class. The instructor should specify the nature of the misconduct, provide evidence if necessary, establish goals for behavioral improvement, and identify specific consequences for failure on the part of the student to improve her disruptive behavior. The driver trainer might consider using a contract or agreement (Figure 3-1) with any students who misbehave in order to provide a concrete and agreed-on plan for behavioral improvement.

Any type of behavior that impedes learning can be considered a barrier to learning for all students. Experienced teachers and educators have developed techniques for dealing with difficult and challenging students. The driver trainer's role in these situations is to get the class back on track and to minimize any negative effects of the behavior or misconduct on the learning of others in the class.

Techniques for presenting information to students and for managing the learning process and classroom behavior are discussed in Chapter 5, "Communication Skills."

ADVISING STUDENTS

Many state regulatory agencies and accrediting bodies require that students have access to academic advisement from members of the faculty, including referral to professional assistance if necessary. In addition, students whose academic progress is unsatisfactory must be counseled and provided with any needed assistance. Driver trainers may be called on to provide information and advice to students on subjects such as employment opportunities, driver's licensing and regulatory requirements, industry trends, industry regulatory requirements, trucking company information, opportunities for career advancement, and continuing education. Instructors need to be prepared to counsel and advise their students on a range of student placement and development topics.

Driver trainers also must be prepared to identify any students having difficulties at home or in their personal lives that affect their academic and skill development progress. The instructor must be prepared to identify problems of a personal nature and to provide a referral to professionals or agencies that are qualified to assist students with specific personal problems. The goal of these referrals is to assist students with any factors that might prevent their educational progress.

Educators should schedule regular sessions with every student to discuss his or her academic progress through the course of study. Such sessions should provide summaries of how the student is progressing in understanding of theoretical concepts and the application of specific skills. Additional areas that can be addressed during these review sessions

This agreement is enetered into between:

Learner: _____

Driver Trainer: _____

Terms of Agreement:

_____

_____

_____

_____

_____

_____

_____

_____

_____

_____

_____

_____

_____

_____

_____

Consequences of failure to comply with the terms of the agreement:

_____

_____

Date and requirements for evaluation of the effectiveness of the agreement:

_____

Learner's Signature _____

Date _____

Educator's Signature _____

Date _____

FIGURE 3-1   Sample Educator-Learner Agreement

include the learner's professionalism, attitude, communication skills, and ability to interact with others. These employable skills are routinely identified by employers as critical to employee success.

A private session provides an ideal opportunity to outline all areas in which the learner is meeting or exceeding performance standards, as well as to identify any areas that need improvement. The driver trainer and the student should agree on a plan of action to improve any areas identified as unsatisfactory.

GUIDELINES FOR HELPING STUDENTS DEVELOP EFFECTIVE STUDY HABITS

The following information is suggested as guidelines that driver trainers can use to help their students establish and maintain effective study habits.

Accept personal responsibility for learning. Be responsible for your own learning by accepting responsibility for your learning process and its outcomes. Taking charge of your own learning allows you to make decisions about your priorities, time, and resource use.

Define personal values. Understand your own personal values and principles, and center your actions around them. Following through on your principles will allow you to develop self-confidence as a learner and teacher.

Set priorities and goals. Decide what is important in order to accomplish your career goals. Develop and assign priorities according to your goals in a logical and sequential manner. Assign resources according to your priorities.

Establish and follow a schedule. Develop a personal schedule that will permit you to meet other obligations as well as your educational priorities. Develop a time budget and an action plan that will help you to accomplish all activities within a set time period. Follow this schedule and designate a set time each day to devote to your studies and skill practice.

Avoid distractions. Acknowledge that you will need a break from time to time. Separate from the place of studying by standing up and taking a physical and mental break from focusing on one subject or activity. Try to avoid distractions that can pull you away from your studies.

USE OF EDUCATIONAL AIDS

Every learning institution should maintain learning resources appropriate and essential for the achievement of the objectives established for each program offered. Learning resources include relevant educational materials such as reference books, periodicals, business manuals, technical references, textbooks, workbooks, audio-visual materials, audio-visual equipment, and other support materials.

RECORDKEEPING

Driver trainers will be responsible for a certain amount of recordkeeping and reporting related to managing a classroom. These requirements are established by the training institution, the state regulatory agency, the accrediting body, and the industry certification organization. To comply with all necessary requirements, educators must know all rules and regulations that apply to the truck-driving industry and to truck-driving training schools.

CONCLUSION

The responsibility for creating a positive classroom environment challenges driver trainers to use their skills as leaders, teachers, supervisors, managers, advisors, disciplinarians, and student advocates. The effective driver trainer learns to balance each of these roles effectively and is rewarded with positive and stimulating experiences with learners in the classroom and in the laboratory setting. Educators must skillfully integrate organizational and communication skills to manage a classroom effectively. They must set an example for their students and develop and display organized work habits to ensure that administrative tasks are completed as effectively as teaching tasks. Educators must ensure that their classrooms are safe environments that promote solid student learning and skill acquisition.

Effective trainers also realize that they must clearly outline their expectations for students and make themselves available to help students address any issues that might present barriers to effective learning. Trainers should address any challenges that arise in the classroom with consistency and fairness. They must listen to their students and strive to maintain student interest and motivation to learn.

Suggested Ideas for Trainer Training Exercises

1. List the factors that can affect the quality of learning.

2. The educator's image, attitude, and actions are often _____ by learners' behavior.

3. Behavioral research indicates that learners will react more _____ and _____ behaviors can be shaped more quickly and permanently through positive means rather than punitive actions.

4. Specific disciplinary techniques selected to address any misconduct should be carefully selected on the basis of _____.

5. The driver trainer's role in a student misconduct situation is to _____ and to _____ on the learning of others in the class.

6. Conduct a role-play exercise between an instructor and student who is misbehaving and discuss the strategies and approach used by the individual playing the instructor to address the misconduct. Create a sample agreement to address this situation.

7. Driver trainers also must be prepared to identify any students having

   _____that affect their academic and skill development

   progress.

8. When an instructor observes that a student is having difficulties of a

   personal nature that affect learning, the instructor must be prepared to

   _____and be prepared to _____that are

   qualified to assist students with specific personal problems.

CHAPTER 4

Presenting Information to Students

OVERVIEW

Presenting information is fundamental to training in any setting. Before presenting information to students, it is important that educators understand as much as possible about who their learners are in order to determine the best methods for presenting information.

For driver trainers, presenting information means using instructional skills such as demonstration, coaching, and storytelling—skills not always used in other training settings. Driver trainer training means addressing new skills and offering a learning opportunity participants may not have encountered before. These topics require that the session leader demonstrate the new skills and provide an opportunity for participants to experience the skills as part of the learning process.

As with the other chapters in this handbook, this chapter is intended as a resource guide for those who create trainer-trainee programs to use in constructing their programs. It contains suggested exercises to use in training driver trainers, objectives that may be useful to the master trainer, and printed materials that may be useful during the instructional process.

Each of the suggested learning exercises will require from 15 to 60 minutes to complete with a group of trainees. Many of the exercises involve demonstration and role-playing since it is critical that adults learn by actually performing the skills they are expected to master. Driver trainers should use trucking examples so that participants have an opportunity to develop a repertoire of ideas and possible situations to use as they begin working with trainees.

There are a variety of ways trainers can present information associated with drivers learning their craft. The particular method used to present information will be determined by both the type of information to be presented and the learning styles of the trainees. Remember, regardless of the methods selected, trainers must set the stage for instruction by:

• Organizing the learning activity.

• Indicating precisely what is expected of the learners during and after completion of the learning exercise.

• Explaining the meaningfulness of the content to trainee job needs.

As discussed in the previous chapter, instructors should vary their strategies for presenting information to stimulate learning. Besides methods for presenting information according to learning style, additional approaches can be used.

Teaching approaches can be varied by matching content with method, by matching learner characteristics with method, and by matching trainer and learner need for variety with method. Often instructors find one method with which they feel comfortable and use only that method. The result is that, after a while, both the instructor and the learner become less attentive to the information in that learning situation and learning rapidly declines.

By varying the methods of presentation and by attending to the steps outlined in these materials, trainers can maintain interest and learning efficiency. Varying the method of presentation is something that trainers must consider before, during, and after each lesson. These methods—demonstration, coaching, and storytelling—have been proved to work best for driver trainers, almost to the exclusion of other methods.

OBJECTIVES

As a result of reading and studying this chapter, you should be able to:

- Understand effective presentation practices for driver-training learners, based on specific understanding of the learners.

- Perform each of the three types of techniques discussed and achieve "good" ratings on at least 80% of factors on the evaluation form.

- Given a sample presentation, trainers will correctly critique each technique used and correctly identify the critical factors.

UNDERSTANDING PRESENTATION PRACTICES

Students judge educators and trainers by what they hear from them. You may be the most efficient and accomplished educator at your institution, but these factors will not serve you well if you cannot communicate your knowledge, skills, and accomplishments to your students. Your success as a driver trainer will depend on your ability to influence the students in your classroom. You will need to present information in a manner that your students can understand. Your students must not only hear you, they must understand and respond to you. To present information effectively, driver trainers need to understand who their learners are and the best methods to present information according to learner needs and backgrounds.

The diverse backgrounds of today's adult learners create even greater challenges for driver trainers to communicate and present information effectively for all enrolled in your class. As previously mentioned, it is important that the effective driver trainer understand cultural and other factors that might affect their students' learning. Educators also need to understand the generational issues brought to the learning process by the various age groups in their classrooms.

Following are steps the effective driver trainer should consider to increase personal awareness of learners and their cultures and backgrounds.

Steps for Increasing Personal Awareness

Maintain clear and open communication. Driver trainers need to create a learning environment that keeps the lines of communication open both among students and between students and the instructor.

Speak clearly and concisely. This is especially important when teaching students whose first language is not your own. It is also important to refrain from using slang, as non-native speakers might not understand your references and may be reluctant to indicate they do not understand.

Attend to nonverbal clues from students. Behaviors and mannerisms change from culture to culture and generation to generation. Be sure to check your understanding and interpretation of a learner's nonverbal messages. Pay close attention to the body language and facial expressions of non-native speakers. They may state or indicate they understand what has been taught when they really don't understand. Ask these students to explain what you said or demonstrated to help ensure comprehension.

If an instructor has established a good rapport with the entire class, other students may help the non-native-speaking students with translations and understanding. In addition, non-native-speaking students may confide in classmates, so it is important to build and use support from the entire class in helping these students understand the materials.

Consider causes of conflict. When misunderstandings occur, consider the possibility that they may be based on cultural or generational differences or perspectives.

Check your acceptance of cultural and generational differences. Understand that your class may contain individuals from different generations or cultures and that people from these groups may have different perspectives, values, and traditions from yours. For example, people from older generations, who did not grow up with today's learning technology, might need extra time when using computer-based learning applications. When presenting information, remember this!

Creating Student Motivational Circumstances

To present information to students effectively, it is important for driver trainers to create a learning atmosphere by which your students can maintain their motivation to learn and excel.

Establish strong personal contact. Make yourself available to your learners in order to discuss any factors that may affect their potential for learning and skill acquisition. Practice effective listening techniques.

Encourage student involvement. Vary presentation techniques to create a stimulating learning environment. Encourage learner participation and questions. Work to keep your students in an active learning mode.

Use examples and illustrations. Do not assume that learners will apply what you are teaching. Support your content with personal examples and illustrations to strengthen the learning objectives. Be prepared to demonstrate how the information you are presenting can apply to the needs of each learner.

Give praise, recognition, and approval. As outlined, adult learners do not appreciate being embarrassed. Take care to avoid criticism in a public situation or in front of peers. Be free with praise and positive recognition.

Encourage personal competition. Students improve greatly when they learn to measure where they are at a given point on the learning continuum. They determine how they can improve their personal performance.

Display enthusiasm and excitement. Your students will follow your lead when you display excitement about a learning topic. In addition to the interest and enthusiasm you show about mastering a particular skill, you should be available for your students before and after class. Use good eye contact with students during your presentation.

Identify long-term benefits and stress internal motives. One of the effective strategies for helping your students acquire new skills is to help them understand how they will benefit in the future from the information or skills you are presenting. If your students can begin to understand how what they learned today will be important to future use and success, they are more likely to retain what they have learned.

Encourage learner support of one another. Encourage your learners to support one another in the learning process. You may not be able to connect with each learner every day, but by encouraging the development of peer support networks in your classes, all students can benefit and obtain the maximum benefit from the education provided.

Offer your learner choices. When offering projects, exercises, or activities for a class, provide choices for your learners so they can complete application techniques of personal value and feel that they are in control of their own learning processes.

When considering how to train, it is important to reflect on the effectiveness of various methods of processing information. According to researchers, the average North American person retains approximately:

- ❏ 10% of what he *reads.*
- ❏ 20% of what he *hears.*
- ❏ 30% of what he *sees.*
- ❏ 40% of what he *sees* and *hears.*

Such a learner is relatively passive; that individual is "receiving information." By contrast, the same individual may remember approximately:

- ❏ 70% of what he *says.*
- ❏ 90% of what he *says* and *does.*

Learner retention is about 80% or more when an answer is given to a question the student has asked. Other students gain from this interaction and information exchange as well. This fact represents a primary reason an instructor should allow students to ask their *entire* questions, no matter how many times the instructor might have heard them before.

These percentages are only approximations, but they indicate where emphasis in training should be placed. Learning can be improved by keeping the following ideas in mind.

❑ People learn faster by seeing and hearing than by hearing alone.

❑ People learn still faster when doing is added to seeing and hearing. It is doing that makes learning permanent. *Adults learn best by doing.*

❑ People tend to remember what they did in training rather than what they were told in training.

Thus, people should be trained for key positions under conditions that are as much like the actual job as possible.

THE DEMONSTRATION TECHNIQUE

What Is Demonstration?

Demonstration is a process in which one person (you, as the instructor, or a trainee) does and explains something in the presence of others to illustrate a point or to show how to perform an activity. The method is especially useful because:

❑ Learners can see what happens.

❑ The action holds learners' attention.

❑ Completion of the activity illustrates and depicts performance standards and procedures.

❑ The demonstration technique reduces the potential time for later trial and error learning.

❑ Actual demonstration can help to illustrate abstract points.

What Are the Advantages of Demonstration, and When Should You Use It?

As a teaching technique, demonstration offers unique advantages and opportunities.

Advantages of demonstration include the following:

❑ It offers information in more than one way and stimulates many senses simultaneously.

❑ It attracts and holds trainee attention.

❑ It ties together theory and application that helps to convince the learner more quickly.

As effective as demonstration is as a method for presenting information, it should not be used as the sole method for teaching a lesson. Effective trainers use demonstration *only* in combination with good explanatory background materials. You should use demonstration to show trainees how to do a task or perform an operation, and to clarify and apply a principle. When note taking is required, the trainer should combine demonstration with the additional training techniques of coaching and mentoring.

How Do You Use Demonstration as a Training Technique?

Demonstration can be time consuming to plan and use, but, it provides a means for immediate application of skill and discussion of relevant information. One caution concerning the use of demonstration is that it can imply simplification of the information being presented.

Demonstration can provide trainees with a real sense of accomplishment as they learn by doing as they practice and report the demonstration.

The rules for using demonstrations are simple, yet critical:

❑ Plan and rehearse the demonstration in advance and in its entirety to make sure that it works, that you have all the necessary materials, and that you can successfully demonstrate the skill as it should be performed on the job.

❑ Set the stage for the demonstration by introducing key concepts and explaining desired learner outcomes.

❑ Always state the purpose or objective of the demonstration and the successful performance expectations.

❑ Be sure you can answer the question of why successful demonstration of this skill is important to the life and work of the trainee.

❑ Make certain the learner can see and hear the activity.

❑ Demonstrate the process and explain it at the same time. First, do the process at full speed, so that learners' initial exposure includes the skill and proficiency completed in real time. After the initial demonstration, slow down and perform the process again, this time at a slower speed.

❑ Ask questions, provide feedback, and register important points throughout the demonstration. Invite learners to ask questions.

❏ Recap the major steps and points at the end as a summary. Also, consider repeating parts of the process, focusing on showing the correct ways to do things. Consider demonstrating the incorrect way to complete tasks if you feel it will help the learning process.

❏ Require learners to replicate the activity, explaining to you (and each other) what they are doing as they perform the activity. Look for signs of confusion, and clarify the process. Coach trainees to correct behavior.

❏ Explain and highlight safety issues, always recapping the most important points.

❏ Prepare and use a *procedure task sheet* so that the trainees can follow the rules and refer to the sheet for future reference. A completed task sheet can serve as a reminder to trainees that they have mastered certain skills, providing an important tangible foundation for self-confidence in their learning abilities.

When Would You Use a Demonstration?

A demonstration produces the best results when trainees need to learn a new skill or procedure. It is easy to have a demonstration ready for a certain time; however, it is more difficult to have the trainees ready for a demonstration.

Usually, new drivers are excited about learning and they are receptive to your suggestions. Virtually any kind of visual behavior or skill can be demonstrated. Just remember to plan the demonstration carefully and work trainees through each step, to review and to have trainees practice each demonstration thoroughly. Also, it is a good idea to have written instruction sheets available for the demonstrated task and to use them for trainees' review as well as a future reference.

Here are some additional points to remember as you plan and utilize demonstration training techniques:

❏ Slow down and perform the procedure correctly. Show how to do the process, *not* how to make mistakes.

❏ Explain the why of each step, and provide key points and cautions.

❏ Demonstrate your own high-quality craftsmanship. (Be thorough and never do a sloppy job before your trainees.)

❏ Give the trainee an opportunity to try out her new skills after a demonstration.

THE COACHING TECHNIQUE

What Is Coaching?

In many work settings, coaching is provided by a senior worker to a more junior worker or trainee and is seen as a system of one-on-one guidance, reinforcement, support, and correction. The use of the coaching technique is often limited to a specific task, such as learning new skills associated with a new job.

Coaching is a very practical and task-focused method of helping trainees develop new skills. It is often a one-way flow of information and demonstration from the coach to the trainee. Coaching can be a dynamic, empowering, and multidirectional exchange for both the coach and the trainee. By understanding how beliefs shape daily experiences, the trainee as well as the coach can be challenged to grow and achieve beyond the limits of simply learning new skill sets on the job.

Coaching creates a positive outcome that is supported by both the coach and the trainee. It avoids the traditional superior and subordinate roles. Instead, coaching provides a supportive environment that encourages growth through:

❑ Increased self-knowledge

❑ Intelligent risk-taking based on informed decision-making

❑ Modeling

❑ Feedback from an experienced source (the coach)

❑ Improved or changed performance

Coaching begins with explanation of the basic concept, related definitions, objectives and expectations. Sometimes, the coach also demonstrates a task, with suggestions. Coaching often provides an opportunity for both the coach and the person being coached to do the following things:

❑ Discover hidden beliefs

❑ Assess self-concepts

❑ Develop positive self-talk and a personal image of success

❑ Discover and expand comfort zones

❑ Develop individual and personalized self-fulfilling prophecies

❑ Create related, measurable goals

What Are Its Advantages, and When Should You Use the Coaching Technique?

In the standard model of coaching as a training technique, the trainee receives most of the benefit. However, coaching also brings benefits to both coach and trainee by providing for:

❏ Improved sense of self-worth

❏ Fulfillment of the coach's own growth and development needs

❏ Opportunity to give back to the system

❏ Renewed interest in one's own life and career (through an exchange of wisdom for energy)

❏ Increased self-awareness

The trainee benefits by learning from a coach who is a seasoned veteran. The coach offers the trainee sound, tested advice, as well as encouragement and emotional or psychological support. By being in a one-on-one coaching relationship with an expert, the trainee learns through the experience of others. Trainees also build a relationship that allows them to continue to talk with the coach about important issues long after the formal coaching process is completed. Additionally, the trainee gains job as well as life skills. She accomplishes this through:

❏ Increased self-esteem and increased self-efficacy

❏ Improved levels of personal goal achievement

❏ Ongoing support and guidance from someone who cares

How Do You Use Coaching as a Training Technique?

The fundamental principles of coaching should be used in almost all interactions with the coaching team. The following example illustrates an appropriate use of the coaching technique and demonstrates how this approach deals with life skills as well as work skills.

A new driver trainee is struggling with a family issue, such as being away from his wife and new baby. The coach can explore and challenge how the trainee's beliefs and fears about being away during a period of adjustment to a new baby affect his work and behaviors with others. By helping the trainee address his fears in a straightforward manner, the trainee can assess his own beliefs and the reality of the situation. The coach also can help the trainee create a support network and communication routine that will help to ease his discomfort about being apart from his wife and child, that will help maintain regular communication, and that will provide assistance in case of a family emergency. This process helps the trainee replace fears with positive action and a more useful belief system.

The coach can help the trainee by using examples of her own experiences during similar circumstances, and then describe how the outcome was successful or a positive learning experience. The coach, by identifying with the trainee, can help the trainee to clarify his own life issues and reach a resolution that will benefit the trainee's life and work skills.

Other strategies that successful coaches use include the following:

- ❑ Listen and encourage the trainee to talk.

- ❑ Ask clarifying questions and rephrase what you hear to ensure that you understand.

- ❑ Offer multiple ideas or suggestions.

- ❑ Deal both with trainee feelings and the content of what he is saying.

- ❑ Explore how feelings are real and affect other phases of life.

- ❑ Be honest in responding.

- ❑ Remember how you would like to be treated in a similar situation.

THE STORYTELLING TECHNIQUE

What Is Storytelling?

A story is a narrative that tells the listener about something that occurred in the *past,* is told in the *present,* and will be remembered and guide behavior in the *future.* Stories can motivate, persuade, and inspire trainees. Further, stories work much better than the usual, straightforward presentation of facts and ideas because trainees can visualize themselves in a story and "see" the action take place in their mind's eye. However, you must select a story that truly exemplifies whatever you may have as a learning objective. The storytelling technique works best when you use a story that actually applies to whatever subject you may be presenting.

Telling a story is like giving your audience a present. Everyone likes to receive presents, and everyone likes to hear stories. Part of the power behind stories is the timing of when they are delivered. To be truly effective, stories have to be delivered at the right time. This takes some talent, because you have to develop the intuitive sense to know when to tell a certain story. In education, this is called a *teachable moment*. Trainers can introduce the story by saying something like, "Let me tell you a story...," or, "That reminds me of a situation I once encountered...."

What Are Its Advantages and When Should You Use Storytelling?

Stories are word pictures. They stimulate memory, and they enable the listener to imagine being in the situation you are describing. They involve both emotions and intellect. Stories work in almost any situation in which knowledge or attitude is key. Based on the information provided on how individuals learn best, you begin to understand how a technique such as storytelling can help trainees absorb knowledge.

How Do You Use Storytelling as a Training Strategy?

Based on available research, there are some proven technologies to improve your ability to use stories effectively. They include the following recommendations:

❏ Do not memorize the story. Tell it in your own words. You might want to learn the story by practicing to tell it to a class.

❏ Make the story short. Usually, short stories are the most powerful.

❏ Use detail. Tell what, where, when, who, how, and the outcome. Emotions, place, and time help make a story real and memorable.

❏ Describe action. Most good stories have a central character who must overcome a problem, with whom listeners can identify. Describe the character, the problem, and necessary actions to reach an outcome. Discuss both what happened and how the central characters felt.

❏ Pause. For any story to be effective, you must secure and continue to keep the listeners' attention by building their anticipation. One of the best ways to do this is to master the art of the dramatic pause.

❏ Be in the story. To tell stories powerfully, you can't stand apart from the story—*you must be part of the story.* You can't simply be a news reporter.

❏ Allow your voice to show emotion. When you tell a personal emotional story, the listener usually becomes riveted. The more you are into your story, the more your listeners will be into the story. *Be into it for maximum effect!*

❏ Make the story fit the audience. Make the story as relevant as possible to your listener. For example, tell a gasoline transporter about gasoline transporters; tell your trainees about other trainees.

Directions: Count the number of circles in each column. Multiply the number of circles in the "Very Good" column by 4; in the "Good" column by 3; in the "Fair" column by 2; and in the "Poor" column by 1. Combine all the totals to determine the Final Total. The Final Total must be at least 48 points to pass.

	Very Good (4)	Good (3)	Fair (2)	Poor (1)	Not Observed (0)
1. Allots enough time for demonstration, practice, and coaching.	Very Good	Good	Fair	Poor	Not Observed
2. Assembles all necessary information, materials, tools, equipment, and supplies.	Very Good	Good	Fair	Poor	Not Observed
3. Explains purpose or objective, context, and outcomes.	Very Good	Good	Fair	Poor	Not Observed
4. States key concepts beforehand.	Very Good	Good	Fair	Poor	Not Observed
5. Sets up so learner can see and hear.	Very Good	Good	Fair	Poor	Not Observed
6. Demonstrates process correctly and at full speed first.	Very Good	Good	Fair	Poor	Not Observed
7. Demonstrates and explains process at 1/3 speed next.	Very Good	Good	Fair	Poor	Not Observed
8. Invites questions and discussion; emphasizes important points.	Very Good	Good	Fair	Poor	Not Observed
9. Stresses "how to" rather than "how *not* to do."	Very Good	Good	Fair	Poor	Not Observed
10. Recaps all major points.	Very Good	Good	Fair	Poor	Not Observed
11. Asks learner to repeat activity, demonstrating and talking about the process.	Very Good	Good	Fair	Poor	Not Observed
12. Coaches by correcting mistakes, showing the "best" way, and reinforcing proper procedure.	Very Good	Good	Fair	Poor	Not Observed
13. Highlights safety factors.	Very Good	Good	Fair	Poor	Not Observed

FIGURE 4-1   *Demonstration and Coaching Rating Sheet*

		Very				Not
14.	Provides a task sheet that includes all correct steps, in proper sequence.	Very Good	Good	Fair	Poor	Not Observed
15.	Asks questions and responds appropriately.	Very Good	Good	Fair	Poor	Not Observed
	Totals	_____	_____	_____	_____	_____
	Grand Total	_____				

Instructor Name: _____

Date: _____

Rater Name: _____

Presentation Title: _____

FIGURE 4-1   (Continued)

CONCLUSION

The diverse backgrounds of today's adult learners create even greater challenges for driver trainers in communicating and presenting information effectively. To present information effectively, it is critical that driver trainers understand the backgrounds of their learners. Driver trainers must also create circumstances and situations by which students can become motivated and maintain their motivation to learn and adapt the information being presented.

Presenting information is fundamental to training in any setting. For driver trainers, presenting information means using instructional skills such as demonstration, coaching, and storytelling—skills not always used in other training settings. This chapter has provided information on each of the presentation techniques and provided opportunities to develop skills in each of these presentation strategies.

Suggested Ideas for Trainer Training Exercises

The following activities are recommended as ways in which to teach critical presentation skills for trainers.

1. Illustrate, discuss, and practice demonstration, using a variety of trucking tasks, such as shifting or backing. What are some additional skills that can be effectively taught using the demonstration technique?

2. Ask each trainer to think about the following situations and develop a story about each that teaches the critical skills and knowledge drivers need in each situation.

   • Emergency at home
   • Running out of money on the road
   • Stopped for a ticket
   • How to deal with a dispatcher asking you to cheat
   • Managing conflict at terminal

3. Role-play how to deal with a number of situations that drivers have encountered, such as:

   • Conflict with another driver.
   • Problems with dispatchers.
   • Problems with a spouse or family member

4. Enact role-playing exercises that demonstrate effective training techniques. For example, a female trainee makes a mistake during her first experience behind the wheel of a rig. Have two trainees role-play the following:

   • The female trainee displays behavior that reflects negative self-talk.
   • How would the role-play coach help her challenge her negative self-talk by exploring her beliefs about her abilities as a woman in a nontraditional field?

- How would the coach help the trainee choose positive self-talk that supported her goal of becoming a licensed load carrier?

- Change roles and have the class instructor play the trainee. A trainee will play the coach.
- Try a different scenario, in which the trainee has positive beliefs and uses positive self-talk to demonstrate how to offer acknowledgment, encouragement, and support.

5. Life inventory exercise. This exercise can incorporate the use of lists, a narrative or journal style of writing, or any other form with which the individual is comfortable, including pictures, sketches—whatever the imagination can create. Ask yourself questions about your life and where you are in it. Look at all parts of your life: your job or career, your relationships (family and friends), your physical and mental health, finances, and so on. Ask and answer questions like these:
   - What in my life am I satisfied with? What am I dissatisfied with?
   - What about my life would I like to change?
   - What would I like to keep or have more of?
   - What do I get really excited about?
   - If my life could be anything I wanted, what would it look like?
   - What would I like to experience in the next year? In 5 years? In 10 years?

6. Hidden beliefs exercise. This exercise should be presented in a classroom setting, but participants must be assured that the outcome can be kept private, unless they choose to share with the group. Encourage honesty. These beliefs are not good or bad. They just are, and should be used as data, not for judgment. The exercise is designed to bring deeply held beliefs to the surface for acknowledgment and examination. Once the beliefs are brought to awareness, individuals can assess, reject, change, or keep them.

What were you taught about yourself as a child? What messages (not necessarily spoken messages) did you get about the following?

Intelligence

Compassion

Gender

Creativity

Sense of humor

Health

Strength

Athletic ability

Other qualities not mentioned

What lessons were you taught about these things?

Money

Love

Marriage

Work

Responsibility

Individuality

Commitment

Tolerance for others

Religion

Men

Women

Other ethnic groups

Families

Taking care of yourself

Other qualities/issues not mentioned

7. Personal qualities list. As you work on this exercise, make sure to avoid all desire to judge or blame yourself or others. This exercise is to help you with a big-picture look at your life and to help get a specific idea of how you might want to change the picture you see.

 • Make a list of your qualities that you see as strengths.

 • Make another list of qualities that you see as weaknesses.

 • List changes that you've already experienced in your life, including those that just seemed to happen to you and those that you sought out.

 • What about the changes was difficult?

 • What was easy?

 • How did you feel after the transition was over?

8. Goal-setting exercise. After reviewing the results of the preceding exercises, choose one thing you would like to change about your life. For example, say you want to become more physically fit through a walking program. You have wanted to be more physically fit for a long time; it has been your New Year's resolution for many years. Now you're ready to set a goal and get moving. Goals must be reviewed regularly and modified, as necessary.

- *Wishes become goals when you write them down.* Write them in ink.

- *Make your goal specific.* Define the timeframe in which you want to achieve your goal. (If you're 40 years old and have never had an exercise program, your fitness goal will not be achieved in three weeks.)

- *Your goal must be quantifiable.* In other words, how will you measure your progress toward becoming more physically fit? If your exercise program calls for walking a mile every day, you could easily use the time it takes you to walk that mile as a quantifiable measure of your increasing physical fitness. As you become more fit, your time to walk the mile will steadily decrease. Another quantifiable measure would be improving your heart rate as you walk faster.

- *Goals must be reasonable.* If you hate to walk, you live in a place where walking is dangerous, and your feet hurt all the time, you probably have chosen an unreasonable goal.

CHAPTER 5

Using Communication Skills

OVERVIEW

Communication skills are perhaps the most important set of skills that a driver trainer needs to ensure that new drivers acquire the skill and knowledge they need to be effective in their profession. This chapter addresses four critical sets of skills that driver trainers should master: listening, understanding nonverbal cues, reinforcing behaviors, and resolving conflict. Each of these skill sets is presented in a series of exercises that you may choose from to help participants learn by doing.

Each of the suggested learning exercises will require from 15 to 60 minutes to complete with a group of trainees. Many of them involve discussion and role-playing since it is critical that adults learn by actually performing communication processes. Many of the exercises use trucking examples so that participants have an opportunity to develop a repertoire of answers and situations to use as they begin working with trainees. As with each chapter, there also are suggested objectives that you, as a master trainer, may choose to employ as you plan your communication skill instruction.

Researchers tell us that the typical person spends up to 80% of his waking hours in a communication process. Of that 80% of time, approximately one-half is spent speaking and approximately one-half is spent listening. Therefore, trainers must master both listening and speaking skills. Through communication, we express emotions and incorporate them into our daily lives. Moreover, through the communication process, people learn how to deal with situations that are part of their lives. It is only through communication that people can ask others to share experiences, look for clarification about things they do not understand, and express feelings in ways that are socially acceptable and productive.

OBJECTIVES

As a result of reading and studying this chapter, you should be able to:

- Identify at least five nonverbal communication clues in a typical communication process and indicate how each reinforces or contradicts the message of the words.

- Correctly indicate two or three techniques that would improve one's listening skills for a particular communication process.

- Identify and demonstrate at least three reinforcement techniques for training.

- Identify and use effective mediation techniques for a potentially difficult situation.

- Identify principles of anger management and demonstrate differences between feeling angry and choosing how to express anger.

LISTENING AND RESPONDING SKILLS

Our society places more emphasis on the spoken side of communication, but if most people are asked to think about those who have had the most significant impact on their life, particularly in the role of a teacher or mentor, chances are they would identify someone who was a good listener. The people who influence us the most are powerful listeners—people who have developed the skill of empathetic listening, who you feel really listen to you and understand you.

Teaching students to listen effectively is one of the most important skills a teacher can impart. Listening is more critical than speaking, and it is much more than just hearing. Listening and hearing are separate and distinct processes. Hearing is an involuntary physical act and is only the first part of listening; it occurs when your ears sense sound waves. There are three other parts to listening that are equally important. There is *interpretation* of what is heard, which leads to understanding, or too often, to misunderstanding. In the *evaluation* stage, you consider the information you've heard and decide how you'll use it. Finally, based on what you heard and how you have evaluated it, you *react and respond* to the other person. It is this combination of hearing, interpreting, evaluating, and reacting that makes up the whole activity called *listening*. Quality listening includes analysis of what is heard and both visual and verbal feedback.

Amazingly, we spend about 80% of our waking hours communicating, and, according to research, at least 45% of that time is spent listening to others. Trainees, in fact, often listen as much as 60% to 70% of the time to others who are providing specific information to them. Even so, very few people have ever received intentional training on listening and as a result, most of us are inefficient listeners. At best, we only remember about 50% of what someone else said, and within 24 hours, most people usually remember less than 25% of what someone said.

Nevertheless, we remember the good listeners we have encountered in our lives, and there are ways to improve our own listening skills. This chapter introduces many of those ways.

First, to master guidelines for *good listening*, there are a few simple rules to follow:

❑ *Stop talking.* You cannot listen effectively if you are talking!

❑ Show you want to listen (check your body language) and remove any distractions.

❑ Focus your complete attention on the speaker.

❑ Try to see the situation from the other person's point of view.

❑ Be patient. Hold your temper. Try not to judge the other person.

❑ When you argue, even if you win, you lose.

❑ Ask questions to show interest and make things clear.

❑ Use active listening skills. For example, paraphrase what you think you have heard in order to verify what you have heard. You can do this beginning with such phrases as, "You seem to be saying..."

❑ *Stop talking.* This is both the first and last guideline because all other listening skill development depends on it. Remember, you have two ears and only one mouth. This fact should provide important guidance: listen twice as much as you talk.

So, how do you know if you are a good listener? One way to begin to assess your listening skills is to take the following quiz. The questions indicate characteristics, including attitudes and actions, that can help you determine if you are a good listener. If you answer each question honestly, based on the information provided, you will find what you need to work on to improve your listening skills. Plan to work on any areas you mark as "No" or "Sometimes."

Are You a Good Listener?

	Yes (75–100% of the time)	Sometimes (50–75% of the time)	No (Less than 50% of the time)
Attitudes			
1. Do you try to see the problem as the speaker sees it?	_____	_____	_____
2. Are you interested in the speaker as a person?	_____	_____	_____
3. Do you listen willingly?	_____	_____	_____
4. Can you remain calm, even though the speaker may be angry and excited?	_____	_____	_____
Actions			
5. Do you take the time to listen?	_____	_____	_____
6. Do you give the speaker your full attention?	_____	_____	_____

	Yes (75–100% of the time)	Sometimes (50–75% of the time)	No (Less than 50% of the time)
7. Do you hear the speaker out, even if you feel the speaker is unorganized and repetitious?	_____	_____	_____
8. Do you withhold judgment until the speaker is finished?	_____	_____	_____
9. Do you try to get the speaker's meaning from his context and content?	_____	_____	_____
10. Do you try to identify the main idea(s)?	_____	_____	_____
11. Do you understand (what the speaker is implying as well as what the speaker is saying)?	_____	_____	_____
12. Do you sense the speaker's underlying feelings?	_____	_____	_____
13. Can you stay focused on what the speaker is saying regardless of distractions?	_____	_____	_____
14. Do you smile, nod, and otherwise encourage the speaker?	_____	_____	_____
15. Do you ask questions (at appropriate times) to be sure that you understand?	_____	_____	_____
16. Can you set aside your biases about the speaker?	_____	_____	_____
17. Do you withhold your response until the speaker is finished saying what he wants to say?	_____	_____	_____
18. Do you look (versus stare) at the speaker most of the time?	_____	_____	_____
19. If the speaker hesitates, do you encourage him to continue?	_____	_____	_____
20. Do you restate the speaker's idea, at appropriate times, to see if you are receiving the speaker's intended message?	_____	_____	_____

DEVELOPING EFFECTIVE LISTENING SKILLS

In addition to working on the items that are effective listening *actions* and *attitudes* from the previous quiz, you can do other things to improve your listening abilities. These skills and actions are presented here. *Remember:* Failure to understand is caused as frequently by poor listening (receiving) as by poor speaking (sending).

❑ **Use and encourage nonverbal signals,** such as:

  ❑ Eye contact.

  ❑ Facial expression.

  ❑ Saying things like "tell me more..." or "hmmmm..."

  ❑ Nodding.

  ❑ Remaining silent.

  ❑ Appropriate posture.

❑ **Establish eye contact.** This is especially important. Think about how frustrating it is to be speaking with someone who fails to return your eye contact. You automatically assume they are not paying attention to you and are not interested in what you are saying.

❑ **Rephrase or paraphrase comments** by restating the main idea or content to:

  ❑ Check or clarify accuracy.

  ❑ Let the speaker know that you understand.

  ❑ Hear yourself say the idea.

  ❑ Encourage more discussion.

❑ **Focus on *key words, main ideas,* and *visual images*** (or examples) to help retain important points. Mentally outline ideas and put them into categories, such as:

  ❑ Similarities and differences.

  ❑ Advantages and disadvantages.

  ❑ A specific chronology.

❑ **Ask questions if what the other person is saying is not clear or complete.** Do not pretend to understand if you do not! The following phrases can help you, as a listener, interject nonjudgmental feedback into a conversation:

❏ **Clarifying.** Get the other person to explain something that was not clear: "I don't understand what you mean," or "Would you repeat that again?"

❏ **Paraphrasing.** Repeat what the other person said, but in your own words: "Let me see if I'm with you. You said…" or "In other words, you think…" or "What I hear you saying is…"

❏ **Summarizing.** Put a number of points into one sentence: "What you have said so far is…" or "As I understand it, your key point is…"

❏ **Take notes**, in list or outline form, if that will help you keep up with the conversation. However, do not concentrate on the notes at the expense of the conversation.

❏ **Avoid bad listening** habits, such as:

❏ Assuming in advance that whatever the speaker says will not be rewarding or interesting.

❏ Mentally criticizing the way the other person speaks instead of listening to the words.

❏ Reacting with such immediate opposition to an idea that you do not even hear the remainder of the speaker's arguments—you are busy thinking of opposing arguments instead of listening to what is being said.

❏ Concentrating on small details and missing the main ideas.

❏ Trying to take notes on everything someone says instead of just the main points.

❏ Pretending to pay attention but thinking about something else.

❏ Remaining silent when the speaker is unclear or incomplete in expressing thoughts.

❏ Tuning out when the speaker uses technical terms that do not mean anything to you.

❏ Letting prejudice against certain words or phrases block your receptivity to the speaker's ideas.

UNDERSTANDING NONVERBAL COMMUNICATION

Nonverbal communication accounts for most of the meaning that passes between two people during interaction. Words account for only

10% of meaning, body language accounts for 55% of meaning, and tone of voice accounts for 45% of meaning. Therefore, nonverbal communication accounts for about 90% of what people understand in a communication process.

When you are interacting with someone else, you pay the most attention to the other person's entire body in the communication process, therefore, you must ensure that your words correspond with the nonverbal information. In addition, you must pay attention to the facial expression, the sound of the other person's voice, and other nonverbal information. This is especially problematic for driver-trainers because either the driver-finisher or the new driver trainee usually is driving as part of the interaction that is occurring. Therefore, it is very difficult to see the other person's eyes or to use the body to reinforce what the words are actually saying. Nevertheless, it is important to recognize and use available nonverbal clues and to incorporate them into your understanding.

Especially important in nonverbal communication are *eyes*. In face-to-face interaction, eyes convey a very large percentage of meaning. They help add emphasis to what is being said. More importantly, eyes convey enthusiasm and sincerity. But let's assume that eyes are rarely involved in the nonverbal communication in the truck.

Other nonverbal communication is almost as important. For example, it is extremely important to notice whether what someone is saying corresponds with the *nonverbal* cues that accompany the message. For example, if the tone of voice and if hand or body gestures reinforce what is being said, it improves the probability that the speaker is sincere. However, if the words say one thing but the nonverbal communication says something entirely different, one should pay particular attention to the contradiction. That does not mean that you should disbelieve the content of the words, but you do need to ask clarifying questions to be sure that you understand the complete message.

Facial expression is a particularly important nonverbal communication factor in the truck-driving setting. Facial expression can convey happiness, anger, sadness, disgust, surprise, fear, and many other emotions. However, facial expression is also the easiest nonverbal communication to fake; actors do it all the time. Therefore, once again, ensure that the words and all the nonverbal cues are in alignment.

Pay attention to *gestures*. Although that might not be appropriate in every situation in the truck, it certainly applies to conversations when the vehicle is not moving, and to conversations with a person sitting in the jump seat. Notice whether or not the other person's body is rigid; also notice if they pay attention to what you're saying by watching or using their hands for gesture, or otherwise indicating enthusiasm in the conversation. Gestures in body language often are more subtle than facial expression or use of the eyes, but they can provide valuable cues to assist you as you try to understand what someone else is saying.

Listen to the person's *tone of voice*. Listen for enthusiasm in a person's voice, particularly as they ask you questions and try to learn more about life on the road. An enthusiastic tone of voice is a very good clue about the extent to which someone is interested in the industry.

The most important thing about nonverbal communication is to look for *consistency*. Nonverbal communication conveys meaning that often provides clues as to what the other person is thinking. Use those clues, in addition to verbal questions, to gain understanding during the communication process. Some nonverbal clues consistently hold particular meaning across large groups. You must also match the nonverbal clue with the verbal message and ask clarifying questions to be sure what the person means. However, pay attention to the following clues and what they might mean:

❑ *Boredom* is often indicated through gestures, such as stretching, looking away from the person who is talking, glancing at one's watch, or folding one's arms and sitting back as a way to imply a lack of receptivity and a break in the conversation. You can try to overcome the message of these gestures by nodding to keep the person involved or by leaning slightly toward them.

❑ *Vocal cues* that demonstrate other types of feelings include a monotone, which may indicate boredom, or very abrupt speech, which can indicate defensiveness or discomfort.

❑ *Body position* also provides some clues, particularly when a person folds his arms or remains very fixed and stiff in the conversation.

REINFORCING BEHAVIOR

In training situations reinforcement techniques provide learners with feedback on the acceptability of their performance and can strengthen desirable performance and minimize or eliminate undesirable performance. The following information outlines specific techniques used to reinforce learning.

When learning theorists hear the word *reinforcement*, they think of techniques that are designed to encourage or discourage behaviors or learning, which is a behavior change. Reinforcing techniques include *reward* (positive reinforcement), *punishment* (negative reinforcement), and *extinguishing* (elimination of a behavior). For example, if a learner answers a question correctly, you could positively reinforce that response by saying something like, "That's very good." This type of positive reinforcement or reward helps the learner to associate the correct answer with the reward or recognition.

In addition, the word *reinforcement* is often used to suggest a technique, such as drill (i.e., any technique designed to strengthen the learner's retention of subject matter or content). For example, if you have a learner who has trouble remembering how many pints are in a quart, you could give that learner 10 problems that require him to apply this concept. Through repetition, the concept of how many pints are in a quart should be mastered. In addition to review and drill techniques, such as audio and videotape replay, summaries and reviews and programmed materials are also used to reinforce content learning.

Positive Reinforcement (Reward)

A *positive reinforcer* is any action by the trainer that encourages the learner to behave in the desired way. Reduced to its most basic form, positive reinforcement (reward) theory states that when a learner performs some act, such as giving a correct answer to the instructor's question, and he is rewarded for it, he is more likely to repeat that act in the future.

The purpose of the positive reinforcement is to reward the correct behavior that took place *immediately* before the reinforcement. The effect is to make that behavior more likely to recur. As the learner repeats the response and is given further rewards, the behavior becomes more firmly established, until it is learned. The more consistently he is rewarded, the more readily the learner will learn.

As an effective trainer, your task is to arrange the training situation so that learners will seek the available rewards as they proceed in their learning. To be effective, the reward that you use to facilitate learning must be something the learner wants or finds pleasant.

Rewards may take the form of special privileges (e.g., being excused from doing a disagreeable assignment) or personal approval (e.g., a smile or a word of encouragement). Often the rewards for learning will come from the reinforcement provided by the learning outcomes. For example, as a learner learns how to back the truck, the finished product and expertise of the learning (backing successfully in many different situations) will be the reward.

Keep in mind, however, that the reinforcement must be something that the learner considers desirable, not necessarily something you think is a reward. Therefore, you must know your students to apply effective reinforcement.

As new subject matter is presented, reinforcements or rewards should be given every time the learner responds in the desired way. After the initial learning, you can give fewer reinforcements.

Research demonstrates that if learning is to take place, the following four guidelines must be followed:

1. The subject matter must be presented to the learner at his own level.

2. The subject matter must be presented in a logical sequence.

3. The learner must know when he is making correct or incorrect responses.

4. Reinforcements must be given as the learner gets closer and closer to the defined goal. This indicates that both you and the learner know what the learning goals are and that the learner should be positively reinforced (rewarded) as he progresses toward the goals.

Positive reinforcement is by far the most useful and effective type of reinforcement to use in teaching. Most learners like being told they are doing a task correctly. The learner who is being positively reinforced in his learning attempts probably will try to repeat the desired motions until he/she has learned the task perfectly.

Reinforcement must take place *after* the learner makes the first contact with new learning, not before. It is also important that the learning be reinforced quickly after it happens—within a few seconds is best. If the learner gives the correct answer, you should immediately say something such as "That's right!" or "Very good thinking!" Even a few minutes' interval between the behavior and the reinforcement can make the reinforcement less effective. In fact, sometimes a reward delayed has the effect of no reward at all.

Positive reinforcement is any action that encourages the learner to behave in the desired way. You have only a limited variety of reinforcing words, expressions, motions, and rewards available to you; therefore, you must learn to use them well. The following lists suggest some of the positive reinforcements that you may use in the course of a lesson. Some of these are tangible rewards, such as prizes and small gifts. Others are intangible rewards, such as expressions of approval, physical expressions, or privileges. Some of these words, phrases, actions, and expressions may seem unnatural or awkward to you, but every instructor will need to develop a repertoire of positive reinforcements that suit his personal teaching style, learner group, and manner of expression. For example, one instructor may say, "That is excellent work," while another will express the same positive reinforcement by exclaiming, "That's dynamite!"

The following are some *spoken words* that you could use as positive reinforcement:

Yes	Correct	Right
Good	Excellent	Terrific
Wonderful	Perfect	Continue
Nice	Beautiful	Remarkable
Okay	Exciting	Exactly
Great	Fine	Super

The following are some *spoken phrases* of approval you could use:

Fine answer.	That's clever.
Go ahead.	That's interesting.
All right.	You perform very well.
Of course.	I'm pleased.
That's true.	Thank you.
How true.	That shows thought.
A good way of putting it.	Very well done.
Keep going.	Well thought out.
Good thinking.	You're doing better
Good response.	You are improving.
Right on!	That is the best yet.

The following are some *nonverbal expressions* of approval you could use:

Looking at the learner

Smiling

Nodding

Grinning

Raising eyebrows

Laughing happily

Shaking the learner's hand

Clapping hands

Signaling okay

Thumbs up

Circling hand through air to encourage the learner to continue

Patting the learner on the back

Moving toward the learner to talk to him

Other *actions* you can take to show your approval include the following:

Asking the learner to show what he is doing

Asking the learner to explain something to others

Using the learner as a model of correct performance

Asking the learner to participate in a demonstration

Giving the learner free time

Awarding the learner points toward a reward

To summarize:

❑ A positive reinforcer is something that increases the probability of the behavior being repeated. It is usually something the learner likes or wants.

❑ Positive reinforcement must follow the desired behavior, not precede it.

❑ The reinforcement should follow the learner's response immediately or very shortly afterward.

❑ At first, the positive reinforcement should be given for every correct response, then less frequently.

❑ The reinforcement must be omitted when the desired behavior does not occur.

Negative Reinforcement (Punishment)

A *negative reinforcer* is any action taken immediately after certain behavior that discourages repetition of the behavior. Negative reinforcement is ordinarily something that the learner does not want or will not like and is considered punishment. Although this definition is not a strictly scientific one, it is useful for the practicing instructor. The purpose of negative reinforcement is to discourage repetition of the behavior that occurred immediately preceding the reinforcement. The effect is to make that behavior less likely to reoccur.

Negative reinforcement may have serious unwanted side effects if used carelessly or insensitively, and you should avoid frequent use of it in instructional settings. There are times, however, when a situation calls for its use. In that case, negative reinforcement should be followed by or coupled with some form of positive reinforcement.

Negative reinforcement must be used very sparingly and carefully. Some signs of disapproval are comments like, "That was dumb," grimacing facial expressions, or ridicule. Such reinforcement techniques can be damaging and have no legitimate place in training.

Negative reinforcement is a form of punishment, and if you become associated with punishment, your effectiveness can be impaired. If you have a good, strong relationship with your learners, however, mild negative reinforcement may spur them to greater effort.

The following list suggests a few negative reinforcers that are sometimes acceptable for training. Some *spoken phrases* of disapproval are as follows:

> That's not exactly correct.
>
> You need to think that through again.
>
> Incorrect.
>
> Unsatisfactory.
>
> You can do better than that.

Some *nonverbal expressions* of disapproval are as follows:

> Looking disappointed
>
> Shaking head
>
> Frowning

Other *actions* you can take to show your disapproval include the following:

> Stopping instruction and quietly staring at the learner
>
> Withdrawing a desired privilege from the learner
>
> Giving a poor assessment to the learner

You have the very difficult task of trying to determine when and how much to reward or punish learners. It is difficult to know what constitutes reward or punishment for each learner. This must be learned from experience and sensitivity to individual learners' reactions. Different cultural and socioeconomic groups respond to different reinforcers, and individuals within groups vary widely in what they perceive to be a reward or a punishment.

Extinguishing Behavior

When a learner's behavior is not followed by any reward or other reinforcement, that behavior is likely to occur less frequently. No reinforcement tends to reduce the behavior to extinction and is, therefore, referred to as *extinguishing*. The most common form of extinguishing is ignoring a behavior.

Extinguishing well-learned behaviors will require time and patience on your part. In fact, the undesirable behavior may actually increase at first as learners try desperately to repeat behaviors they previously found reinforcing. Another hazard is that the learner who is ignored may view this as a punishment and may stop trying to participate at all.

RESOLVING CONFLICTS EFFECTIVELY

Conflict resolution and anger management are closely related concepts. In order to mediate and resolve conflict between others, and to be better at understanding our own expressions of anger, it is helpful to explore some basic principles of each concept. Both conflict resolution and anger management rely heavily on listening and responding skills. The listening and responding skills previously presented are the building blocks for conflict resolution and anger management.

Managing Anger

The state of anger is often confused with aggressive expressions of anger. Anger is not a primary emotion, as many of us have been taught, but a state of physical readiness—nothing more and nothing less. When we are angry we are prepared to act. When we are angry, the following things happen in our bodies: adrenaline production increases, more sugar is released into our bloodstream, our hearts beat faster, our blood pressure rises, and the pupils of our eyes open wide. Anger is a biological response designed to ensure survival. Anger is preparedness and power.

We are most powerful when angry, which is why anger can be so dangerous. People who use their anger destructively with themselves or others can cause a lot of damage and pain. Anger is a normal, physical reaction to some stimulant in our environment. For many people, it has become an almost automatic reaction to frustration, emotional hurt, or fear.

Contrary to some beliefs, anger is not blaming, sarcasm, violence, vindictiveness, viciousness, aggression, sulking, manipulation, scapegoating, yelling, or cursing. These things are *learned expressions of anger*. Anger and the expression of anger are separate things and must be considered separately. Anger and aggression are often thought of as related, or even as the same thing, but they are decidedly different.

There is no person or event on the face of the earth that can "make" you angry. When something happens to hurt, frustrate, or frighten you, you might respond with what feels like automatic anger. The true relationship between the event and your reaction is a response learned by reinforcement over many years and events. No matter how automatic your angry reaction feels, the truth is that you, and only you, are in control. Only you can choose how to react to the hurt, frustration, or fear you feel.

How does one move from a conditioned response of aggression or an explosive outburst when angry, if that is all you have ever known?

❑ First, accept the truth that all our behaviors are things we choose.

❑ Second, monitor beliefs about any given thing that "makes" you angry. Is your belief legitimate? Is it reasonable? Is it a relic from the past? Does the belief work in your favor or against you?

❑ Third, and most important, get in touch with the feelings behind the anger and your chosen expression of it.

When your dispatcher gets that look on his face that reminds you of a former domineering and sarcastic boss, do you get angry and react by storming out of the office because you feel helpless, as you did when you worked for that boss?

When we come to accept responsibility for our behaviors, we can experience a great sense of freedom from our perception of others' control over us. By choosing to be in charge of our own feelings and responses, we get to decide how we act and react. We are free to behave as we choose! No one else can ever be blamed for our feelings or given credit for our actions.

Once you have explored these issues within yourself and examined your own behavior, you can begin to help others explore their expressions of anger. By following the three steps suggested above, you can challenge others to take more responsibility for their behavior, be more aware of their feelings, and begin to make different choices about their expressions of anger.

Preventing Conflicts

One of the best ways to resolve conflicts is to prevent them. As a trainer, you can overcome many potential sources of conflict through discussion and consideration of issues that typically cause a problem in the truck. Often, your initial conversations with a new trainee can cover some sensitive topics and help to forestall future problems. Pay particular attention to issues you know cause conflict. Discuss the situation. Work out rules and agreements about how you will treat each other in various situations. For example, discuss the following issues that typically arise when people are driving and traveling together:

- ❑ Music choices
- ❑ Climate control
- ❑ Smoking
- ❑ How to wake each other up
- ❑ Topics of conversation and appropriate language
- ❑ Home visits and calling home
- ❑ Sleeping and hotel stops
- ❑ When the trainee can ask questions
- ❑ Recreation
- ❑ Handling anger and disagreements—when you agree to disagree

Try to reach agreement on how you will deal with each other as you begin. Then work on and talk about issues as they arise.

Conflict Resolution through Mediation

The goal of conflict resolution is to help each individual involved in the conflict develop a mutually agreeable plan for action through mediation. Conflict resolution is not a silver bullet—not all conflict resolution works, and not all disputes can be mediated. Conflict resolution is almost always a compromise in which each individual agrees to do something more to the liking of the other. The compromise, or resolution, ***must*** be mutually agreeable to all individuals involved in the conflict. As a mediator of conflict, you must always begin with this goal in mind.

Mediation is useful in situations in which two parties have not been able to communicate openly or clearly and those in which conflicting expectations exist or issues of territory arise. For example, let's say another driver and a dock worker have had a problem. If tempers have flared to the point of violence by the time you arrive at the scene, all disputants must be separated, given instruction to take a cooling-off time-out (15 minutes, for instance) and asked to report back at a specific time to begin to resolve the dispute.

Begin the mediation process with two basic assumptions: (1) Conflict can be constructive and, when viewed positively, is seen as a sign of a need for change, and (2) angry people are hurt people. The following are basic rules of mediation:

1. To avoid being seen as taking sides or having one or the other party fabricate or exaggerate details, the mediator must bring all the disputants together in the same place at the same time. Separate interviews are more time consuming and send a subtle signal that the mediator is responsible for the ultimate outcome of the conflict.

2. The mediator must maintain control of the situation by keeping everyone focused on the issues, by staying in the "here and now," and by stopping any bickering or provocative behavior immediately. All these things are attempts by the disputants to express their anger and be heard. Assure everyone that by following the rules, everyone will be heard.

3. Establish rapport with the disputants by using the techniques of active listening, repeating back what you've heard, clarifying to confirm your understanding, and probing through the use of open-ended questions that begin with the words *who, what, when, where, under what circumstances,* and *in what way.* Using these words to begin questions allows you to avoid "leading" the disputants in their answers. Use of these words ensures that you get specific information about behaviors and avoid assumptions about the thoughts behind behaviors. Questions beginning with the word *why* are closed questions and limit the information available to the mediator.

   ❑ Avoid closed questions.

   ❑ Use silence as a tool to make disputants feel heard.

   ❑ Avoid breaking a silence to end your own discomfort.

❑ Finally, use review and summary techniques to help disputants feel you are actively engaged and to understand their positions.

❑ Briefly summarize the main points and ask each disputant if there are things to be added.

❑ Stay with each individual until he is satisfied that you have all the information and you understand it.

4. It is extremely important that you stay neutral. If you agree or sympathize with any disputant, you lose control as well as credibility.

5. Use the 80:20 ratio to determine who does the talking. The mediator must stay in the 20% category, the disputants in the 80%.

6. Stay in your role of interviewer. Answer questions that relate to process only. Use review and summary to avoid answering other questions.

7. Mediate one problem at a time by keeping everyone on topic. Draw discussion back to the stated problem by saying things like, "Today we have agreed to discuss..." You will surely lose control by allowing a disputant to confuse or diffuse the current topic of mediation.

8. Remain impartial at all costs. If you are seen as taking sides, disputants shut down and become defensive, hostile, and resistant to successful resolution.

9. Accept each disputant's view on face value. There is no right or wrong point of view.

10. Avoid making suggestions. Rules of social etiquette do not apply in conflict resolution, except those that relate to being respectful of others' rights.

11. Encourage disputants to express their feelings. All disputes have an emotional content, and usually hostility and anger are easier to express than hurt. Acknowledging feelings helps build rapport.

At this point you have completed the information-gathering phase of conflict resolution and are now prepared to move to support the disputants in reaching resolution. Begin by asking for specific suggestions from the disputants. "I don't know" is not acceptable as an answer. Press for specific change by always staying at the "doing" level. "Doing" things are easier to measure. Check all suggestions with the other disputants by repeating the suggestion. If the suggestion is accepted, you are on the way to an action plan. If the other disputant rejects the suggestion, immediately ask that person for his suggestion for a solution. Disputants learn quickly that if they reject an idea, they must then come up with the next idea. If the original idea seems reasonable, the rejecting disputant begins to look unreasonable. Remember the rule to stay neutral! Agreeing with or criticizing a suggestion at this point could destroy the whole process. Additionally, if you make a suggestion at this point, you become responsible for the outcome of the process.

Accept small changes and continue in a back-and-forth process until all disputants agree on some specific change. When agreement is reached, review and sum up the plan. Check with all disputants for agreement, and then acknowledge the plan verbally by saying something like, "This looks like a workable plan." Next, put the plan in writing. A written plan adds to the significance and importance of the agreement. The written plan should be clear and specific, using language as simply as possible. The plan must be signed by everyone, including you. Plan for a follow-up meeting, if applicable, that you include in the written agreement. Finally, reward the success! Support the way the process has worked. Focus on what worked, not on what failed.

CONCLUSION

Communication skills are perhaps the most important set of skills that a driver trainer can master and use to ensure that new drivers acquire the necessary skill and knowledge to be effective in their chosen profession. In this chapter, four critical sets of skills that driver trainers should master have been addressed: listening, understanding nonverbal cues, reinforcing behaviors, and resolving conflict.

Suggested Ideas for Trainer Training Exercises

1. Pick a television show and practice good listening. Perform a mental checklist of the listening skills used in the text and practice taking notes on a 10-minute segment. When you are finished, recap the main ideas and think of the questions that you might ask as a follow-up.

2. Watch people in a conversation, and make notes about the use of nonverbal cues that either reinforce or contradict the words that are being said. Discuss these with the larger group.

3. Practice using nonverbal communication in some kind of setting by playing some kind of pantomime activity.

4. Divide participants into pairs and give each group two sets of five identical dominoes. Arrange the pairs sitting back-to-back. Ask person 1 to arrange his dominoes into a pattern and then instruct his partner (who cannot see the pattern) how to replicate the pattern. Person 2 may neither speak nor look at the other's pattern. Repeat the process, but allow questions. Discuss the effectiveness of instructions and their differences, as well as the value of questions.

5. *The anger thermometer.* As an in-class exercise, have everyone draw a vertical line down the middle of an $8\frac{1}{2} \times 11$ sheet of paper, then horizontal lines at regular intervals along the vertical line. This is the anger thermometer. Beginning at the lowest line, have each person write a word that represents the beginning of his personal progression toward anger, such as *annoyed, uncomfortable,* and *rubbed the wrong way.* Have each

person write a word or words on the very top line of the scale that represents how he feels at his angriest, such as *enraged*, *violent*, and *beside myself*. Working from the bottom to the top, each person should fill in as many words as he can that describe how his feelings change as he gets angrier and angrier. Discuss the exercise in class. Listen for and challenge dysfunctional beliefs, such as, "That guy always makes me so mad when he gets in my face." Encourage people to accept responsibility for their reactions to others and to get to the feelings behind the reaction.

6. *Role-play situations.* Present scenarios in which conflict often arises in trucking. Think about past experiences between dispatchers and drivers when tempers flared. Using the mediation skills outlined in the conflict resolution section, have class members practice resolving issues. By playing both roles, class members relate more easily to the perspectives of others and develop more tolerance for the other's position. Break the group into small groups and rotate roles until everyone has an opportunity to play all roles. Follow the role-play with a class discussion. Support changes in perspective and insights about the positions of those normally thought of as adversaries. Collect situations to role-play from the group.

7. *Anger-aggression values exercise.* Before we can change our expressions of anger, we must learn more about the values we hold about anger and how we express it. By answering the following questions, we begin to clarify what our present expressions of anger mean to us. The answers to these questions can be kept private, but everyone should be encouraged to participate honestly. By honestly looking at the answers we give to these questions, we can begin to gain more control over our feelings and our expressions of anger. Anyone open to sharing in a group discussion should be supported in the effort.

   • What, if anything, do I enjoy about getting angry?
   • When I am angry, do I really want to be in control? Or do I enjoy "acting in the moment" because it frees me of responsibility?

- What do I believe or think about explosive and impulsive acts when I am angry? When others are angry?
- When I am angry with people, what outcome am I looking for?
- What do I believe about tolerance of others, mine for them and theirs for me?
- Do I ever knowingly deny my anger? When? With whom? Under what circumstances?
- How do I feel about sulking and pouting? Do I ignore it in myself but criticize it in others?
- What are my beliefs about hatred? About hostility?
- What do I want my expression of anger to be?
- What am I willing to do to be more in control of my expressions of anger?

8. Practice interviewing each other to assess interests, skills, and individual differences. Assume that this is a conversation with a new driver trainee and the trainer has an opportunity to learn about important potential areas of conflict, such as music preferences, recreational activities, smoking, and climate control. Also practice working out the rules that the driver and trainee will use in the truck.

9. Pass whispered verbal sentences in a circle and compare the original and final sentence. Discuss how the words and meaning changed over time.

10. View videos on how to deal with conflict. Discuss techniques and practice skills.

11. Present and discuss typical situations that generate anger, such as a trainee who repeats the same mistake many times. Brainstorm ideas on how to deal with the situation. Develop a list of situations that generate anger and discuss ways to handle each situation.

12. Illustrate and discuss how and why a trainer must explain why to a trainee. Develop a series of issues and answers trainers can expect to deal with.

13. Reinforcement knowledge. The following items check your comprehension of the material about reinforcement techniques. Each of the nine items requires a short essay-type response. Please explain fully but briefly, and make sure you respond to all parts of each item.

A. Explain why a particular reward may reinforce learning with one learner, but not with another.

B. In what way is positive reinforcement related to motivation?

C. How does reinforcement theory apply to instruction?

D. Name three techniques for encouraging and discouraging behavior, and give one example of each.

E. Accept or reject the statement, "Negative reinforcers should be used as often as positive reinforcers in instructional settings." Support your position.

F. Why must an instructor give reinforcement immediately after a desired behavior?

G. Assume that an instructor made the following statement about a given behavior: "That's good, do it again!" What type of reinforcement is being used, and what assumptions can you generally make about the behavior?

H. Negative reinforcement is somewhat similar to extinguishing. Explain the similarity.

I. How can strengthening techniques be used to facilitate the learning of content?

Answers to Reinforcement Questions

Compare your written responses to the self-check items with the model answers given below. Your responses need not exactly duplicate the model responses, but you should have covered the same major points.

A. Just as there are cultural and socioeconomic differences between groups, there also are differences between people within each group. As a result, people tend to respond differently to different reinforcements. John may be very pleased to have his project displayed by the instructor as an example of good planning and construction, whereas Larry might become very hostile because the peers in his group do not approve of group members setting an example.

B. Reinforcement occurs after a learning act and is intended to motivate a learner to want to repeat that act. Therefore, positive reinforcement is a method of motivation that is used after a desirable behavior has occurred.

C. A widely accepted premise in reinforcement theory is the notion that when behavior changes, learning occurs. That is, learner behavior is learned, it doesn't just happen. Further, it is accepted that behavior can be predicted and modified. This theory has direct application in the classroom. By applying reinforcement principles, the instructor can reward good behavior, encouraging learners to repeat it again—and punish or ignore undesirable behavior, thereby discouraging learners from repeating the behavior.

D. *Positive reinforcement* is any action taken by the instructor to encourage a learner to repeat desired behavior. For example, after a learner gives a correct response to a question posed by the instructor, the instructor immediately responds, "That's very good thinking, John!"

Negative reinforcement is any action taken by the instructor to discourage undesirable behavior. For example, a learner (Ed) was absent for several days in order to take a vacation in Florida with his parents. The instructor had sent along all the assignments so that Ed would not fall behind. When he returned to class, the activity for that day was to take a quiz on all the material Ed had missed. Ed handed in a blank answer sheet. The instructor marked an "F" on the paper and assigned Ed extra homework until he had caught up on the missed work.

Extinguishing is the planned absence of any reinforcement. For example, the instructor established a rule in class that each learner wishing to participate in class discussion must first get permission. Tim was eager to answer a question and started to answer before he got permission. The instructor ignored him and recognized another learner who had requested permission to speak.

E. Positive reinforcement should be applied much more often than negative reinforcement, because negative reinforcers can be damaging if not used with caution. Generally speaking, negative reinforcement is something learners do not desire. In some cases, depending on the sensitivity of the learner, negative reinforcement can have serious side effects. Therefore, for the majority of instructional situations, the statement would not be acceptable.

F. Reinforcement is most effective when it is applied immediately after learning—within a few seconds is best. Since this is a case of desired behavior, the instructor wants to have the best assurance that the behavior will be repeated again. The best assurance is to apply the reinforcement immediately after the desired behavior. In this way, the learner is most likely to associate the reinforcement with the desirable behavior.

G. In this situation, the instructor positively reinforced the learner's behavior. In addition, one can make the following assumptions about the behavior:

- The behavior exhibited by the learner was desirable.
- The probability of the learner repeating the behavior has been increased.
- The learner will receive frequent reinforcement for the behavior if the instructor wants it to be learned.

H. Negative reinforcement and extinguishing are similar in that they are both used to discourage undesirable behavior.

I. Among other possible uses, strengthening techniques (e.g., summaries, drills, and audiotape replays) are used to provide the learner with opportunities for repeating the same instruction or content until he has mastered it.

CHAPTER 6

Planning Instruction

OVERVIEW

Successful instructors must know what they want to accomplish, how they want to accomplish it, and when they want to accomplish it. The lesson plan organizes that information. It is a road map for teaching and works much the same way as a map works for planning a driving trip.

This chapter is intended to acquaint instructors with the process of planning instruction so as to ensure that content is covered and that trainees derive maximum benefits from their efforts.

When delivering classroom and laboratory-based instruction, it is likely that driver trainers are following a lesson plan prepared by the safety or training departments. Therefore, instructors must understand several important principles associated with how lesson plans are constructed and used. Driver trainers may create their own outlines for truck-based instruction and will need guidelines to produce lesson plans in a useful and thorough fashion.

As with the other chapters in this handbook, this chapter is intended as a resource guide for those who create trainer-trainee programs to use in constructing their programs. Therefore, they contain suggestions about appropriate exercises to use in training driver trainers, they contain objectives that may be useful to the master trainer, and they include printed materials that may be useful during the instructional process.

OBJECTIVES

As a result of reading and studying this chapter, the reader will be able to:

- Describe the importance and function of lesson plans.

- Understand how to use lesson plans to teach new driver trainees.

- Be able to develop useful lesson plans and outlines for training.

USING LESSON PLANS

A properly developed program of study will result in the orderly and systematic process of education that ensures that students progress satisfactorily through the course and achieve all the established educational objectives. Preliminary analysis of the programs, course objectives and goals, lesson plans, and scheduling must be done before teaching can take place.

Prior to the development of lesson plans, a curriculum development process will be conducted that includes identifying the course content, defining learning outcomes, organizing the material, and developing a course outline. Once these steps are completed, lesson plans will be developed that include methods for evaluating student learning and outcomes.

The core curriculum for truck driver training may be prescribed by a state regulatory agency, an accrediting agency, or an industry certification body. In the truck-driving training industry, these agencies have created curriculum standards based on the skills, knowledge, tasks, and duties of entry-level truckers, as explained and rated by truck drivers. These standards also incorporate curriculum recommendations by the U.S. Department of Transportation. In addition to these curriculum standards, many truck driving schools also use the input of an Industry Advisory Council to ensure that the latest information related to industry trends and requirements is also included in school courses.

Many companies and schools provide their driver trainers with lesson plans that have been prepared by the safety department, the department of human resources, or others within the organization. These lesson plans are intended as a guide for instructors to use when teaching a particular part of a lesson.

To be effective, the lesson plan should have sufficient information so that any instructor in the organization who has content knowledge can be handed that lesson plan, walk into an instructional setting, and deliver the lesson effectively. The lesson plan should be clearly written, flexible, and individualized for conducting each lesson. Each lesson should be based on the needs, interests, and abilities of students and should address the goals, needs, and style of the individual driver trainer. Your success as an effective driver trainer will depend in large part on your ability to plan and present your subject matter effectively. You need to know exactly what you are going to teach and how you are going to teach it.

An effective driver trainer will want to make use of the many tools and resources available for improving the quality of education offered in the classroom. A well-prepared lesson plan is one of these tools. It requires the educator to carefully consider the selection of the subject matter, the procedures, and the preparation of tests to assess student progress. As student profiles and industry standards and performance requirements change, lesson plans will need to be revised and updated to ensure that the most current information is being delivered.

A partial list of the advantages achieved through daily lesson plans includes these:

❏ School owners and driver trainers can be confident that course and lesson objectives will be met.

❏ Learners receive the benefit of effective teaching strategies, such as being able to connect new material to information previously learned and having information presented in a logical and practical sequence.

❏ Educators are prepared with appropriate review questions and answers.

❏ Learners receive the benefit of appropriate summarization and review.

❏ Learners know exactly what is expected of them and have the benefit of follow-up assignments to ensure that they achieve and can demonstrate the necessary level of competence.

❏ Driver trainers have the opportunity to consider a variety of learning styles when developing each lesson plan, which ensures that more students benefit from training.

❏ Educators are required to gather needed teaching materials and aids to provide a high-quality presentation.

❏ The confidence of the educator is increased because she is prepared and knows that the material will be presented in an organized manner.

Lesson plans vary in form and content, depending on the subject matter and style of the educator. Several parts of the lesson plan are especially important and require trainer attention in planning activities. These parts may be labeled under different headings within the lesson plan, but the parts typically include the following items.

❏ *Learner objectives.* Learner objectives are statements of what the trainee is expected to know and be able to perform at the conclusion of training. When objectives are well written, they include not only the action that someone is expected to undertake, but also how well the person is expected to master the subject area and the conditions under which the typical performance is expected to occur. Learner objectives may include a focus on the acquisition of knowledge and skills or the development of attitudes. These objectives are important not only for the instructor to understand as she begins the lesson, but also for trainees to recognize and understand so that they realize how the content applies immediately to work. Learners should have a clear understanding of what they will be able to accomplish or know on completion of the lesson and assigned practice.

❏ *Materials and time.* The materials and time statement in a lesson plan tells the instructor what supplies, materials, and length of time are required to teach a specific lesson appropriately. It allows the instructor to determine which lessons can fit into which time periods and which props are necessary to teach a particular lesson.

❏ *Setting.* The educator will want to indicate where the presentation will take place. If the lesson is strictly a theory presentation, the classroom should be listed. If it is a practical lesson with a demonstration, the laboratory setting (range, road, or both) should be designated.

❏ *Bridge.* The bridge is the technique suggested in the lesson plan to be used by an instructor to tie lesson content to the life of the trainee. That is, the bridge explains why the content is important in the immediate life of the trainee; usually, it connects new information to old information, and new information to some of the trainee's job requirements. In addition, the bridge serves as an attention-getter in the typical lesson.

❏ *Teaching process.* The teaching process typically includes three or four critical elements. Often, it is an outline or statement of steps that details the specific content of the lesson, the suggested teaching sequence, recommended instructional technique, and key points. This section of the lesson plan may be the most important part because it shows what content must be addressed during the lesson. Moreover, it indicates specific opportunities for the instructor to change methods of presentation, thereby focusing attention on the content and the key points.

❏ *Practice.* The practice section of the lesson plan shows when trainees have an opportunity to learn by doing. Remember that adults master material most efficiently when they learn by doing and that lesson plans for driver trainees should include a large segment of hands-on learning. The practice section of the lesson plan indicates when, under what circumstances, how long, and to what level of competence trainees will work on specific skills to master and demonstrate their mastery.

❏ *Evaluation.* The evaluation section of the lesson plan indicates how and when trainees will be assessed to determine if they have mastered the content. The evaluation may consist of a quiz, a discussion of content, or a formal test.

As you examine the lesson plans prepared by other instructors or other officials at your company or school, look for these broad themes or parts. Within each of these themes is contained the information that you need to be successful in using a lesson plan that someone else has prepared for you.

PREPARING A LESSON PLAN

Effective lesson plans usually are well thought out and well constructed. They require time to prepare. Because of the time involved, some instructors do not write out lesson plans—they say they "don't have time." Therefore, they attempt instruction without a plan, and that's usually a mistake. An instructor should no more consider teaching without any type of a plan than a driver should accept a load to move without a map. No matter how experienced an individual may be, both endeavors will turn out to be substandard performances. Consider that commercial airline pilots with millions of miles flown still use a written checklist during takeoff and landing. The written list is the only way they can ensure that they will not forget a key step or operation. So too, a lesson plan is the only way an instructor can ensure that she will not forget a key step or operation.

Realistically, instructors will not always have time to write a formal lesson plan. In that type of situation, a lesson outline should be constructed. Any written plan is better than no plan at all. It does not have to be a typed masterpiece to be effective. If your time is limited, jot down a few highlights, sequence the information, use the information, and make notes to improve it for next time. And remember, just as road maps are updated annually, lesson plans need revision and adjustment as you teach.

When preparing a lesson plan, several planning steps are especially critical.

❑ First, decide on the objectives of the lesson. Find the objective by answering the question, "What must the trainee know and be able to do at the conclusion of the lesson?" Write out the objectives on your plan *and* explain them to the trainee as training begins. Direct instruction is helpful, so that the trainee learns some specific skill or knowledge during instruction.

❑ Consider how to organize the material. Always start at the beginning, not in the middle. Typically effective ways to organize information include going from simple to complicated, from known (and often similar) to unknown, and from past to present. Remember to start where the trainee is, not where you are as the instructor.

❑ Find a way to get the trainee's attention. Often, the attention-getter can be the bridge to trainee interest. The bridge means that you find a way to make the material of the lesson important to the life of the trainee. Answer the questions, "Why is this important to me as a new driver?" and "How does the new information relate to what the trainee already knows?"

❑ Select instructional methods or ways to present information. Use various methods as a way to maintain interest. Three methods are especially useful to driver-trainers: demonstration, storytelling, and coaching and mentoring.

❑ Provide a way for the trainee to learn by doing. Remember that adults learn best this way. Practice not only makes perfect but also is essential to learning in the first place.

Also consider how to determine if the trainee learned the information. Assessment methods might include an oral or written test as well as a performance test.

ORGANIZING INFORMATION INTO A CONSISTENT FRAMEWORK

Once you have considered the content and issues, commit the information to a lesson plan. Figure 6-1 presents a formal outline structure of a lesson plan used in training in other crafts. Fill in the information, and you have a simple but effective lesson plan. Suggested Format 1 is especially appropriate for a classroom or laboratory lesson. It also can be used for the truck, but suggested Format 2 (Figure 6-2) is designed for easier use on the road. Please remember that you are expected to plan for and deliver instruction not just on vehicle operation skills, but also on the critical topics of trip planning, time management, customer service, money management, dealing with family emergencies, dealing with law enforcement situations, and problem solving.

Day _____   Time day begins _____   Time day ends _____

Topic _____

Objectives (What will trainee know and be able to do at end of training?)	**Teaching Aids and Equipment** (What do you need to teach the lesson?)
Bridge (How will you relate the topic and its importance to the life of the trainee to get attention?)	**Evaluation** (How will you evaluate learning? How will trainees practice use of information?)

FIGURE 6-1   Lesson Plan Format I

Presentation

Key Points and Outline of Instruction	Time	Methods/Activities and Materials

FIGURE 6-1   (Continued)

Planning and Report:

Fleet Manager _____   Trainee Start Date _____

Truck Number _____   Trainee Release Date _____

Trainee Name _____   Trainee Signature _____

Trainer Name _____   Trainer Signature _____

Ratings Scoring: Needs Improvement = 1 Average = 3 Excellent = 5		Plan Days of Instruction
Continuing Responsibilities		
1. Making morning check calls	1 2 3 4 5	1 2 3 4 5
2. Backing skills	1 2 3 4 5	1 2 3 4 5
3. Pretrip planning and routing	1 2 3 4 5	1 2 3 4 5
4. Daily log skills	1 2 3 4 5	1 2 3 4 5
5. Tarping and tie-down procedures	1 2 3 4 5	1 2 3 4 5
6. Employee and customer relations skills	1 2 3 4 5	1 2 3 4 5
Assuming Responsibilities		
1. Working with dispatch and fleet management on L and E calls on mobile communications	1 2 3 4 5	1 2 3 4 5
2. Plan most economical routes and fuel stops	1 2 3 4 5	1 2 3 4 5
3. Operation of equipment in heavy traffic	1 2 3 4 5	1 2 3 4 5
4. Plan a second route	1 2 3 4 5	1 2 3 4 5
5. Team and solo logging	1 2 3 4 5	1 2 3 4 5
General Knowledge		
1. Understand policies, procedures, and importance of check calls and dispatch communications	1 2 3 4 5	1 2 3 4 5
2. Understand the role of the cargo claims department and procedures involved in reporting a claim	1 2 3 4 5	1 2 3 4 5
3. Understand the role of the safety department	1 2 3 4 5	1 2 3 4 5
4. Professional attitude and appearance	1 2 3 4 5	1 2 3 4 5
5. Money management	1 2 3 4 5	1 2 3 4 5
6. Understanding mobile communications system	1 2 3 4 5	1 2 3 4 5

FIGURE 6-2   Lesson Plan Format 2

CONCLUSION

A properly developed program of study will result in the orderly and systematic process of education that ensures that students progress satisfactorily through the course and achieve all the established educational objectives. Preliminary analysis of the programs, course objectives and goals, lesson plans, and scheduling must be done before teaching can take place.

As you examine the lesson plans, whether they are prepared by you or by other instructors or officials at your company or school, it is important to ensure that certain required elements are addressed in order for lesson plans to be effective.

Suggested Ideas for Trainer Training Exercises

Since the primary focus of this chapter is to help a driver trainer understand and follow a lesson plan and construct an outline for using truck-based instruction, several activities seem particularly useful. These activities can be done independently by each driver trainer. More often, they are done with driver trainers working in pairs to critique and create new lesson plans. Each activity is designed to take between 15 and 30 minutes to complete.

1. Present, illustrate, and discuss several lesson plans used in your company's driver training activity. Discuss lesson plans from classroom-based, laboratory-based, and range- or road-based instruction. Demonstrate and discuss how objectives are presented to trainees in advance of instruction. In addition, discuss the sequence of activities, the time period set aside for instruction, the methods of presentation, the types of practice in which the trainee is expected to engage, and how to assess the acquisition of skills and knowledge. These topics are part of most lesson plans and usually are found in lesson plans prepared by training or safety departments. This activity also provides an opportunity to illustrate the difference between lesson plans used in classroom instruction and the outline more often used in truck-based instruction. Note that classroom-based instruction typically contains a much more elaborate lesson plan, whereas truck-based instruction usually indicates key points that are expected to be covered within a day or two.

2. Read and discuss the chapter, including using lesson plans, preparing a lesson plan, and organizing information into a consistent framework. Point out that these materials are especially concerned with truck-based instruction.

3. Divide instructors into pairs and give each pair a sample lesson plan provided by the company. Ask the instructors to critique each lesson plan. Is it complete? Is the plan sufficient to allow any trainer who is given the

lesson plan to execute it efficiently and ensure that the trainee masters the information? Ask instructors to suggest ways to make the lesson plan more efficient. Ask instructors what they consider the most challenging task required of them should they undertake teaching the sample lesson plan.

4. Using Lesson Plan Format 1 (Figure 6-1), ask pairs of instructors to create a truck-based lesson plan on one subject that they might use in their teaching practice. Among the subjects of particular interest are dealing with an emergency from home, planning a trip, keeping a log book in conformance with company and federal regulations, managing money on the road, and dealing with personal hygiene.

CHAPTER 7

Assessing Student Learning

OVERVIEW

Testing and assessment are integral parts of teaching. They provide evidence that a trainee has learned the skills and knowledge that compose the course content. In addition, they give an instructor feedback about the value of instruction and can suggest the need for additional training to ensure the trainee's mastery of content.

The purpose of this chapter is not to prepare driver trainers to construct tests. The chapter acquaints driver trainers with the purpose of tests, improves their skills and confidence in administering tests provided to them for use with trainees, and enables them to distinguish useful from ineffective tests.

Consider the suggested exercises as well as the written text. Not every company or school will want to use all the materials; some companies may disallow driver trainers from testing altogether. Others may require the driver trainer not only to test but also to create some of the assessments.

OBJECTIVES

As a result of reading and studying this chapter, the reader will be able to:

- Describe how testing fits into a driver-trainer program.

- Explain and demonstrate how to administer tests provided by the training department.

- Given a set of test items or a checklist, the trainer will critique the items.

INTRODUCTION

It is essential that driver trainers measure the knowledge and skill acquired by their students. Testing, measurement, and evaluation are all part of the ongoing learning cycle. Evaluation begins by clearly defining a set of performance objectives based on the content that is taught. Any assessments conducted of student learning should include assessment methods appropriate to the teaching or learning situation. Tests should measure important information students have been taught.

Testing as evaluation is a tool, not a weapon. Many learners dread any kind of test. The pressure to do a good job and the fear of failure is a heavy burden on all trainees.

Many instructors wish they could do away with testing, too. Constructing tests, administering tests, grading, and recording grades can be a tedious job. Testing will be less of a burden, however, if you know exactly why you are testing and what kind of test to use for what purpose.

TESTING PURPOSES

Ask any student why teachers give tests and the answer will be immediate—so that they can give grades. True enough—but that's only one of the reasons. Tests serve other purposes.

Test to Find Out What Has Been Learned

The most important reason to give a test is to find out what has been learned. You may be presenting a great deal of information to your trainees, but if they are not learning it, everyone is wasting time.

You can be pretty sure that your trainees are learning something from you, but unless you test, you may not know what or how much they are learning.

You are not the only one who needs to keep track of progress—trainees themselves need to know how they are doing. They need reassurance that they are mastering the material they are devoting so much time to learning. And, some may need the jolt they get when they find out that their test results lag behind what they should be.

Test to Find Out If Your Teaching Is Working

Testing reveals both how well trainees have learned and how well you have taught them. As a driver trainer you ought to be very interested in how effective your teaching is.

If your trainee does badly on a test, you cannot assume that he is lazy and stupid, even if that may be the case. It may be that you did not teach as effectively as you should have. You can use testing to assess your quality as an instructor.

If the results of a test are poor, it may not be easy to determine immediately what needs improvement—the learning, the teaching, or the test. Perhaps all three areas need improvement, but you should analyze your teaching to see if you need to make some changes. If a trainee does poorly in one particular area, it may be that your teaching on that particular point was not clear enough.

Test to Motivate

When did you do most of your review and studying when you were in school? Chances are that it was usually before a scheduled test. Most trainees study regular lessons just enough to keep up, but if they know they have a big test coming up, they bear down and study. A test is a great motivator. Scheduling tests at intervals will get trainees to review the material you have covered regularly.

Test to Assign Progress Scores, Grades, Licenses, and Certification

Finally, tests are given to assign scores or grades. That's a fact of life. Tests must be planned carefully to make sure that every test is a fair way to assign grades. For licensing and certification of skills, testing is a requirement.

TEST TAKING STRATEGIES

It is important that educators help their students understand how to prepare for testing, especially for high-stakes testing, such as that required for licensure. Much of the preparation for a test begins with development of effective study habits and time-management techniques. Driver trainers should encourage students to prepare mentally and physically for test taking. This includes getting plenty of rest the evening before a test and expecting and being prepared to deal with the test anxiety that everyone experiences.

The more students understand about the type of test they will be taking, the more prepared they will be. Taking practice tests to prepare for actual tests is a good way to help students understand what to expect in the testing environment. Students should be counseled to take deep breaths and try to relax during testing. They should read all directions through carefully and ask questions about any parts they do not understand. It is helpful to skim the entire test before beginning and try to develop a time budget that will allow ample time for each section. Answering the easiest questions first will reserve needed time to attend to the more difficult questions. Answering questions for which students are sure they know the answers will help to build a more self-confident attitude to support the trainee during the remainder of the test. Students should answer as many questions as possible and, for questions of which they are unsure, guess or estimate an answer. On finishing the test, students should look over the entire test to be sure it is completed. Effective driver trainers will help to coach their students through the test-taking process.

PARTS OF A TEST

Testing must take place at appropriate intervals throughout the instructional process. To be most useful, testing should be a standard activity that occurs after completing a unit of instruction.

The decisions test developers must make to evaluate acquired knowledge and skills include determining *the type of test to give* and the *content of specific test items*. For driver-training instruction purposes, two types of test items are recommended: (1) multiple-choice, and (2) checklists. The usefulness of each type of test item varies. The key to a useful test is clear performance objectives based on course content and test items written to the objectives.

What is a performance objective? A performance objective is a concise statement that describes what the trainee must know and be able to do when he completes a lesson or training session. It describes an intended outcome in terms of the trainee's performance. Performance objectives contain several parts:

- ❑ **Task.** The task is the performance the trainee will be able to demonstrate at the end of the instruction.

- ❑ **Condition.** The condition includes what materials, tools, or manuals he may or may not use in demonstrating the performance.

- ❑ **Standard.** The standard describes the minimum acceptable performance.

DEVELOPING PERFORMANCE OBJECTIVES

Performance objectives assist the trainer in determining what should be taught, how it should be taught, and how to evaluate the instruction. Performance objectives benefit the following persons in these ways:

- ❑ **Trainers.** Well-written performance objectives clearly define the expected results of the trainer's efforts. The performance objectives identify where to place the instructional emphasis. Objectives prevent the trainer from teaching only those skills that he feels trainees should know or that he feels prepared to teach. The objectives provide the trainer with a road map for reaching training goals.

- ❑ **Trainees.** Trainees need to know the purposes and expectations of the instruction. Clearly stated objectives at the beginning of each instructional session explain what is expected of them. Objectives also can be excellent motivational tools.

- ❑ **Training supervisors and other interested persons.** Objectives can be used to communicate training program content and standards. Objectives frequently are used to evaluate or review the content of the training program and to provide a clear picture of program content and level.

Objectives help to ensure that trainees are tested only on important information and only on information that was taught.

In addition to performance objectives, effective tests have several other important pieces of information that trainers need to recognize and use. For example, there should be a set of instructions—both to the trainee and to the trainer—about how to take and use the tests. For the trainee, the directions indicate how to answer questions, the value of each question, and how much the value of the total test is worth. For the trainer, the directions are instructions for how to administer the test, grade it, and report the findings to administrative bodies within the organization and to trainees. Effective tests offer both sets of directions and even include examples for trainees.

Among the information that should be provided to trainees is which material the test covers and the proportion of the final grade the test represents for a particular unit or course. Trainees should be told whether an opportunity for retesting is available. Test instructions can also indicate how the trainee will receive feedback on the quality of his performance.

Feedback from the instructor or the test administrator is important both to correct errors that the trainee may have made in answering the question and as a type of positive reinforcement, to encourage the trainee to put forth a good effort in training. In addition, feedback helps to build the bond between the instructor and the trainee by providing an opportunity for discussion of issues and for emotional support.

TYPES OF TEST ITEMS

Several types of test items are available to assess trainee knowledge. It is best to avoid true/false tests and to use multiple-choice or performance tests.

Multiple-Choice Items

Multiple-choice items offer great flexibility for assessing all types of performance outcomes because you as the instructor control the background information, the alternative answers, and the level of performance necessary to answer the question guessed.

Multiple-choice items are made up of two parts. The first part is called the stem. The *stem* is written as a question or statement. It provides the background information for the test item and specifies the type of operation the trainee must use to answer the item correctly.

The second part of the item is the *set of alternative answers*, one of which is either the correct answer or the best answer for the problem presented. The other alternative answers are called *distracters*. In the following example, the stem and the alternatives are identified.

Example

Stem

Which of the following types of nonverbal communication usually convey the most information?

Alternatives

❑ Voice tone

❑ Eye contact

❑ Hand gestures

❑ Posture

The major advantages of multiple-choice items are flexibility and scoring ease. The major disadvantages are (1) the amount of time it takes to write good questions and (2) the fact that occasionally, the answer can be correctly guessed.

Following these rules to write or choose multiple-choice items will simplify item construction and decrease the likelihood of correct guesses.

Rule 1: Design each item to measure an important learning outcome. Ignore unimportant or irrelevant information. Use the behavioral objectives to determine the content areas of the items.

Rule 2: Present only one clearly formulated problem or question in the stem. Explain exactly what is being asked and include necessary information to answer the question. Avoid unnecessary information. The stem should be understandable without looking at the alternatives.

Example

Poor Stem

While everyone processes information mentally at a different rate, what is the average number of words most people can process?

Better Stem

About how many words per minute can the average person's mind process?

Rule 3: Construct four or five alternative answers for each stem. Be sure that only one alternative is correct or represents the best possible answer.

Rule 4: When possible, avoid negatively stated items, either in stems or alternatives. If you use negative construction, emphasize the negative construction by underlining, boldface, or capitalization.

Example

Poor Construction

Which of the following behaviors is not a bad listening habit?

Better Construction

Which of the following behaviors is NOT a bad listening habit?

Rule 5: Write each alternative so that it is grammatically consistent with the item's stem. Also, try to make each alternative approximately the same length as all other alternatives.

Example

Poor Construction

When offering feedback to a speaker after receiving a communication, you should:

- ❑ Describe what you saw and think you heard so you deal with the facts.

- ❑ Focus on behavior and individual personality.

- ❑ Focus on judgment and evaluation.

Better Construction

When offering feedback to a speaker after receiving a communication, you should do what?

- ❑ Focus on description and observation.

- ❑ Focus on behavior and individual personality.

- ❑ Focus on judgment and evaluation.

Rule 6: Randomly use each alternative position for correct answers in approximately equal numbers. Too frequently, test makers avoid the first and last positions while concentrating on the middle position. Therefore, use approximately equal amounts of options for a, b, c, and d, or 1, 2, 3, and 4 for the correct answer among the alternatives.

Rule 7: Develop each item so that it is independent of other items. If items depend on other background information, make sure each is identified clearly in the directions.

Rule 8: In general, avoid using choices of "all of the above" and "one of the above." Both alternatives contribute to guessing or can be used as clues.

Rule 9: Make all alternative answers plausible and avoid clues that may permit elimination of incorrect answers or guessing of the correct answer. Exercise care so that you avoid (a) using key words in the stem and alternatives, (b) stating the best or correct answer in formal or textbook-like language, (c) stating the best or correct answer in greater detail, and (d) using two responses with the same meaning.

Rule 10: Improve the alternatives. Refine the wording of alternatives, being sure to (a) state each in the language of the learner, (b) use common misconceptions or errors, and (c) make each alternative similar in wording, length, emphasis, and kind.

CHECKLISTS AND RATING SCALES

Checklists are useful for evaluating simulations and work samples. Focused specifically on procedures rather than products, checklists direct an observer's attention to critical aspects of performance, such as necessary skills and knowledge. They offer the observer a yes/no or *scaled alternative* for deciding whether the specified aspect of performance in question was displayed. The observer notes the occurrence of the desired performance at the appropriate spot on the checklist as the trainee works. The checklist is scored by adding the number of checks and comparing both the sum and the specific items checked against the performance.

The limitations of the checklist are that (1) sometimes it fails to indicate how well or thoroughly the aspect of performance was enacted if it is only yes/no; (2) it is subject to the individual judgment of each observer; and (3) it is limited to skills and knowledge that have observable characteristics.

The following rules will help you develop checklists and rating scales:

- ❑ Refer to the task-analysis "do" column and determine if a performance test item is needed to measure the trainee's ability to perform the critical "do" skill. If it is, use the following guidelines to develop the checklist.

- ❑ Write a description of the skill that is to be performed.

- ❑ Write directions to trainees in language that will be understood by them. Expectations must be clear.

- ❑ If a case situation is needed, write it.

- ❑ List or get the tools, materials, and equipment that will be needed to perform the test.

- ❑ Include all safety procedures and note special precautions.

- ❑ Write directions for the examiner.

- ❑ If time is a critical standard, include the estimated time that should be allowed for performing the test.

- ❑ Construct a performance checklist of tasks or activities. Ensure sequence and description. Use it to evaluate performance.

The checklist should specify whether all criteria on the checklist are necessary for the student to pass the objectives or if only certain items on the checklist are critical. If so, these should be identified. The list section of these materials introduces a new rating scale for a performance test the laborers are using. Figure 7-1 shows an example of a checklist appropriate for driver finishing training.

Rating: 5 = Excellent; 0 = Failing

Trip Planning

Map reading	N/A 0 I 2 3 4 5
Tolls	N/A 0 I 2 3 4 5
Calculating hours to run, arrival and departure times, and breaks	N/A 0 I 2 3 4 5
Factoring in company procedures/policies (fuel, maintenance, etc.)	N/A 0 I 2 3 4 5
Other: Fuel usage (how to calculate)	N/A 0 I 2 3 4 5

NOTE: If driver runs out of fuel, driver pays for service!

Canadian Paperwork	N/A 0 I 2 3 4 5

Sweep & Pull Trailer	N/A 0 I 2 3 4 5

P.T.A. for Single Driver	N/A 0 I 2 3 4 5

Other Skills (some may not be observed; those not observed should be simulated): Accident procedures (should be simulated)

Accident procedures (should be simulated)	N/A 0 I 2 3 4 5
How did student/co-driver react?	N/A 0 I 2 3 4 5
Did the student/co-driver log correctly?	N/A 0 I 2 3 4 5
Did the student/co-driver complete all necessary paperwork in a timely manner?	N/A 0 I 2 3 4 5
Loading procedures	N/A 0 I 2 3 4 5
Loading	N/A 0 I 2 3 4 5
Sealing/locking trailer	N/A 0 I 2 3 4 5
Calculating axle weights	N/A 0 I 2 3 4 5
Use of sliding tandems	N/A 0 I 2 3 4 5
Loaded call	N/A 0 I 2 3 4 5
Proper scaling procedures (Does student/co-driver know how to drive across scales and deal with state officials?)	N/A 0 I 2 3 4 5
Department of Transportation inspections	N/A 0 I 2 3 4 5

FIGURE 7-I Professional Driving School Student Proficiency Report

CONCLUSION

Testing and assessment are integral parts of teaching. They provide evidence that a trainee has learned the skills and knowledge that compose the course content. They supply an instructor with feedback about the value of instruction, and can suggest additional training needed to ensure mastery of content by the trainee.

Suggested Ideas for Trainer Training Exercises

You can use these exercises to help teach the skills and knowledge associated with assessment and testing. Any of the suggested exercises can take from 15 to 60 minutes to complete. All driver-trainer training programs should pay some attention to assessment and testing, but the amount of time each program devotes to assessment is entirely up to the program.

1. Discuss with the participating instructors why testing is used in training. Also discuss problems associated with testing and how to overcome them, especially test anxiety and trainee fear of the unknown. Recognize that instructors and trainees are likely to have had unhappy test experiences.

2. Collect and critique a series of test items from various types of tests. Include driver's license exams, math tests, and other sorts of assessments. Use the rules for writing multiple-choice questions and checklists associated with these materials as the background for providing the critique. Discuss any problems that instructors working in pairs can find in the test questions that you provide them.

3. Practice administering tests using samples provided by the safety or training department, particularly performance assessments and checklists. Ask instructors to role-play the parts of instructor and trainee, and discuss the kinds of feedback the instructor would provide.

4. Role-play providing feedback to trainees about testing results. Work with instructors to discuss the different types of reactions they have encountered with trainees and help instructors deal with those reactions.

CHAPTER 8

Professional Development

OVERVIEW

This section addresses the professional development issues that effective driver trainers need to consider in preparing for employment in the truck driving training field. By adopting a positive attitude and creating role models toward continuous career and professional development, educators will inspire and train learners to become highly skilled, competent professionals and, in some cases, may even inspire some of their students to become educators of the future.

OBJECTIVES

As a result of reading and studying this chapter, you should be able to:

- Prepare for employment as a driver trainer.
- Understand the reasons for selecting this career field.
- Develop long-range career goals as a driver trainer.
- Identify the type of organization to work for.
- Prepare effective resumes for the successful job search.
- Prepare for the job interview as a driver trainer.

PREPARING FOR EMPLOYMENT

Your primary goal as a result of your choice to become a driver trainer is to be able to obtain gainful employment in a field that you enjoy. In addition to the technical skills necessary to instruct students in truck driving, a driver trainer is expected to have specific interpersonal, communication, and teaching skills. As a result of your driver-trainer training, you have obtained a great deal of knowledge about how an effective driver-training organization should operate. That knowledge will aid you in your search for the right type of organization or school in which to use these skills. It is also important that you take a personal inventory of your interests, personal characteristics, skills, personality, and goals in order to select the type of driver-trainer situation that will best meet your needs.

As previously mentioned, the continuing demand for truck drivers and the individuals who train them will result in many opportunities for driver trainers. However, it is important to research the various teaching options available before you select an organization in which to work.

There are a number of questions you should ask before you begin.

- ❑ What do you really want out of a career in education?

- ❑ Do you want to concentrate on teaching specific skills in the truck-driving training industry?

- ❑ What type of organization would you like to teach in?

- ❑ What are your strongest teaching skills, and how do you wish to use them?

- ❑ What personal qualities do you bring with you to any employment situation?

Driver trainers provide important role models for their students. Your students will be able to detect whether you are satisfied with your career choice. Your ability to help your students with their professional development and achievement of a satisfying position in the truck-driving industry is especially important, given the high turnover rate among truck drivers. Although some of this turnover has been attributed to the demands and lifestyle issues of long-distance truck driving, some of it can also be attributed to the lack of career advancement opportunities for truck drivers. It is important that driver trainers inform their students about the realities of this profession and help equip them to take responsibility for their own career development. The driver trainer should be prepared to model this behavior for her students.

RESUME DEVELOPMENT

A resume is used to summarize your education and work experience and to tell potential employers at a glance about your achievements and accomplishments. A resume conveys the first impression a potential employer develops about you as an applicant, so it is important that your resume capture your achievements and display them in an organized, professional manner that makes the reader want to meet you. Your resume should help an employer determine how you have accomplished what your resume states, as well as whether you will be a good fit for their organization and can meet their needs for a driver trainer.

The average time a potential employer spends scanning resumes to determine who should be granted an interview is about 20 seconds, so it is critical to make the most of this brief period. You must market yourself in a manner that will separate you from other applicants. Always focus on your achievements and do not make the mistake of simply detailing the responsibilities of past jobs—any applicant can do that. Accomplishment statements should always enlarge on your basic duties and responsibilities and quantify information, where possible. Before preparing these statements for your resume, you might ask yourself:

- ❑ How many students were in my classes on average?

- ❑ How many students did I supervise on the range and on the road?

- ❑ What percentage of my students successfully graduated from their course of study?

- ❑ What percentage of my students learned to drive a commercial vehicle safely?

- ❑ What percentage of my students learned to inspect vehicles to ensure safety of operation?

- ❑ What percentage of my students can successfully demonstrate knowledge of the laws pertaining to the operation of a commercial vehicle?

- ❑ What percentage of my students can complete appropriate paperwork correctly?

- ❑ What percentage of my students can successfully demonstrate an ability to plan trips and routes, including managing loads and weight distribution?

- ❑ What percentage of my students can communicate effectively with peers, customers, and supervisors?

These types of questions will help you to develop accomplishment statements that will generate potential employer interest and can help you to create a resume that will set you apart from other applicants.

IDENTIFYING THE POTENTIAL SCHOOL OR WORKPLACE

The staff and training organization or school you are considering applying to should be well connected to employers in the trucking industry. The quality of this connection can be determined by how many employers have demonstrated an interest in finding and hiring the school's graduates. The school staff should maintain files on each trucking employer who contacts them to find trained truck drivers. From these files, students can learn about different companies before interviewing with them. Students preparing to graduate from a high-quality training program should be required to research prospective trucking company customers, to understand what these companies expect in an employee, and to research various factors important to applicants, such as company pay scales and benefits.

It is important to visit any school or training organization where you are considering applying for a driver-trainer position. When you visit the school, pay close attention to the type of image the school presents, the level of professionalism of staff, the satisfaction level of current students, the quality of customer service, the quality and appearance of classrooms and equipment, and educator work areas.

EMPLOYMENT

After you have completed your course of study and visited schools where you are interested in working, you are ready to pursue employment as a driver trainer. The next step is to contact the institutions where you are interested in teaching and send a resume and cover letter requesting an interview.

Even if the institution where you are interested in working does not have an opening and the owner, manager, or personnel or safety director does not want to schedule an interview, you might ask for the opportunity to conduct an informational interview. This will give you important background information in the event of an opening while impressing those who make hiring decisions with your initiative.

INTERVIEW PREPARATION

1. **Employment Application.** Make a list of items that are typically requested on an employment application:

   ❑ Social security number

   ❑ Driver's license numbers, endorsements, certifications

❏ Name, location, and years completed of schools

❏ Physical record information

❏ Driving record information

❏ Names, addresses, and telephone numbers of former employers

❏ Names and telephone numbers of personal references

2. **Personal Appearance.** First impressions are important in applying for a position. Dress in appropriate clothing and make sure that your clothes are clean, well-fitting, and comfortable. Your physical appearance should be neat and clean.

3. **Needed Items.** Bring a copy of your resume with you, as well as a list of any additional relevant facts and figures about your past employment.

4. **Answers to Anticipated Questions.** Certain questions are typically asked during an interview. Think about how you would answer the following questions in an interview setting:

   a. What did you like best about your driver training?

   b. What do you like best about teaching?

   c. In which teaching areas do you feel you are strongest and weakest?

   d. Are you a team player?

   e. Do you consider yourself a flexible person? Explain.

   f. What are your career goals?

   g. What experience did you have as a driver before pursuing a career as a driver trainer?

   h. Are there any obstacles that would keep you from full-time employment?

   i. What assets do you bring to this organization?

   j. What are your strongest communication skills and why?

   k. Describe how you handle challenging students. Give examples.

   l. If you are given a lesson plan to teach a specific unit, describe how you would approach delivering that lesson.

   m. Describe the impact that teachers have had on your learning.

   n. How do you see your job with this organization fitting into your career plans?

Additional points:

❏ Learn all you can about the company you are interviewing with before the interview.

❏ Be prepared to ask questions about the company that you were not able to answer as a result of your research about the company, such as information about various company policies and chances for advancement.

❏ Always be punctual for the interview. If you are unsure of the location, be sure to find it beforehand.

❏ Always be polite, friendly, and courteous. Do not smoke or chew gum.

❏ Expect some nervousness—that is normal. Try not to fidget, and concentrate on appearing confident, creating a positive impression, and projecting a positive attitude.

❏ Speak clearly. The interview must be able to hear and understand you, especially if you are expected to be heard by your students.

❏ Answers all questions succinctly and honestly. Think your answers through carefully before speaking.

❏ Never criticize former employers.

❏ Try to keep your focus on why you are the best candidate for the job.

❏ Ask questions about the company and the performance expectations for the job.

❏ Thank the interviewer for his or her time. Ask about the timeline for making a decision and when you can expect to be informed of the interview results.

CONCLUSION

This chapter addresses the professional development issues that effective driver trainers need to consider in preparing for employment in the truck-driving training field. By adopting a positive attitude and creating role models for continuous career and professional development, educators will inspire and train learners to become highly skilled, competent professionals. In some cases, trainers may even inspire some of their students to become educators of the future.

Suggested Ideas for Trainer Training Exercises

1. It is important that driver trainers take a personal inventory of their
   _____ in order to select the type of driver-
   trainer situation that will best meet their needs.

2. List the five questions you should ask yourself before you consider the
   teaching or driver-training options available to you.

3. Why is it important to list accomplishments on your resume instead of
   job requirements or responsibilities?

4. Write five statements of accomplishment you could put on your
   professional resume as a result of training truck-driving students.

5. What type of employer research should you expect your driver trainee graduates to conduct before selecting a place of employment?

6. What characteristics should you consider and observe when you visit a school where you are interested in employment as a driver trainer?

SUMMARY AND CONCLUSION

Effective and successful driver trainers evolve as a result of their own interests, motivation, commitment, energies, and persistence. Successful educators begin perfecting their technical skills, enrolling in a school to develop specific teaching and training skills, completing the course of study, and obtaining employment in their chosen discipline. Many driver trainers work for some time in their chosen field before entering training to become an educator. Successful driver trainers often feel a "calling" to the art and science of teaching and to molding future safe, professional drivers. These individuals also feel a vested interest in maintaining a solid image of the trucking industry. It is important to equip yourself with the necessary foundation skills in order to ensure you are prepared when an opportunity presents itself to use the training skills you have developed. To create your future success as a driver trainer, it is important to imagine the type of position you believe will meet your needs and work toward that vision. Plan for your future success, prepare yourself to meet any challenges, and you will develop the self-confidence necessary to realize your dreams.

Classes and lessons are the most meaningful when they have been planned, practiced, and refined as needed. It has been said that "good teaching is one-fourth preparation and three-fourths theater." Any successful actor will tell you that good acting requires a great deal of preparation, just like successful teaching. In a class your learners are there to be taught, to interact, and to participate in their own learning process. At the conclusion of the class, learners leave, often having been entertained, but almost always with greater knowledge or skill than they had before they came to class. As in acting, preparing to teach takes time. Each step of that preparation and presentation contributes to the educator's effectiveness and the outcomes of the learners. Teaching is more than knowing the subject matter; it includes preparation, practice, and delivery. Teaching well means knowing and designing learning situations that meet your students' needs.

As a driver trainer, you are an important part of the foundation that makes and maintains this important industry. Without high-quality educators and driver trainers, the industry would not be able to maintain its role in our economy. Best of luck with your future and your part in this industry's future!

Index